D1562949

THE BOUNDARY ELEMENT METHOD FOR ENGINEERS

THE BOUNDARY ELEMENT METHOD FOR ENGINEERS

C. A. Brebbia, *Reader in Computational Engineering,*
Department of Civil Engineering,
University of Southampton

PENTECH PRESS
London : Plymouth

First published 1978
by Pentech Press Limited
Estover Road, Plymouth
Devon PL6 7PZ

Second revised edition, 1980
Reprinted with corrections, 1984

ISBN 0 7273 0205 1

British Library Cataloguing in Publication Data

Brebbia, Carlos Alberto
The boundary element method for engineers. –
2nd rev. ed.
1. Boundary value problems
2. Engineering mathematics
I. Title

515'.353 TA347.B69
ISBN 0-7273-0205-1

Set by Mid-County Press, 2a Merivale Road, Putney, SW15
Printed in Great Britain by Billing & Sons Ltd.,
Worcester.

Contents

Preface

The Boundary Element Method is a technique which offers important advantages over 'domain' type solutions, such as finite elements and finite differences. One of the most interesting features of the method is the much smaller systems of equations and considerable reduction in the data required to run a problem. In addition the numerical accuracy of boundary elements is generally greater than that of finite elements. These advantages are more marked in two- and three-dimensional problems. The method is also well suited to problem solving with infinite domains such as those frequently occurring in soil mechanics, hydraulics, stress analysis etc. for which the classical domain methods are unsuitable.

The classical boundary solution is formulated in terms of influence functions, frequently found in the literature under the general title of 'boundary integral' methods. This method is well known and will not be discussed in this book, mainly because the weighted residual approach used here is more powerful and relates the boundary methods to more classical engineering techniques; all approximate methods can then be interpreted as particular applications of weighted residuals techniques.

The term 'boundary elements' originated within the Department of Civil Engineering at Southampton University. It is used to indicate the method whereby the external surface of a domain is divided into a series of elements over which the functions under consideration can vary in different ways, in much the same manner as in finite elements. This capability is important as, in the past, integral equation type formulations were generally restricted to constant sources assumed to be concentrated at a series of points on the external surface of the body.

This book presents the boundary element method in a simple fashion using computer programs written in FORTRAN, to help the beginner to understand the basic principles of the method. The applications of the method to potential problems and two-dimensional elasticity are discussed in detail, together with the way in which functions of different orders can be used to form the boundary elements on the external surface of the body.

The aim of this book is to introduce the reader to the fundamentals of boundary elements in such a way that he will then be able to solve more complex problems on his own.

I am particularly indebted to my friend and colleague José Dominguez for his help with the computer programming and for his valuable criticism and suggestions during the preparation of the manuscript.

C. A. Brebbia

Preface

1

Introduction

1.1 BASIC IDEAS

Engineers and physical scientists are nowadays well conversed with methods such as finite differences or finite elements. These techniques discretize the domain of the problem under consideration into a number of elements or cells. The governing equations of the problem are then approximated over the region by functions which fully or partially satisfy the boundary conditions. These methods together with other techniques that are applied on the domain will be called 'domain' methods.

Another possibility is to use approximating functions that satisfy the governing equations in the domain but not the boundary conditions. These techniques are called boundary methods and for a number of reasons they have developed slowly up to the present time. They are now being re-examined, mainly because they offer an elegant and economic alternative to the 'domain' methods. They can also be combined with the latter to obtain a better representation of the boundary conditions in a finite element or finite difference program. It is well known that finite element results, for instance, may be very inaccurate due to the difficulty of representing properly the boundary conditions in problems such as those with infinite domains. For these problems it is possible to propose a subdivision of the region of interest into finite elements and to use boundary elements to better approximate boundary conditions such as radiation. (The boundary elements represent the boundary between the finite element region and the infinite domain.) Another important advantage of the boundary methods is that they are usually able to represent regions of stress concentration in a better manner than finite elements, although this will generally depend on the type of approximating function used.

Boundary methods can be of as many different types as 'domain' methods, ranging from simple techniques such as the so called 'indirect' method to the more versatile 'direct' techniques. Both techniques will be explained in more detail in what follows but it is interesting to point out that they can be interpreted as different weighted residual formulations. Weighted residuals have been recently used to classify and systematize the domain methods, and we shall now show that they can also be used for boundary solutions. This approach appears to be the most appropriate way of looking at the approximate techniques in order to simplify their presentation and understand the relationship between the methods.

1.2 BOUNDARY SOLUTIONS AND OTHER METHODS

The least sophisticated boundary technique is the so called 'indirect' method of analysis. In its simplest form it starts with a solution that satisfies the governing equations in the domain but which has some unknown coefficients. These coefficients are then determined by enforcing the boundary conditions at a number of points or sub-regions. This is a crude technique but it does not detract from the fact that it can give good results in many practical applications. This is specially true when the solution used is the fundamental solution of the governing equations. Some of these 'fundamental' solutions are shown in Chapters 3 to 5 of this book, and they represent the solution over an infinite domain for a unit source applied at one point.

The indirect technique has been used for many years due to its simplicity. It is interesting to note that it can be interpreted (see Chapter 6) as a particular type of weighted residual technique.

Direct methods on the other hand, are generally presented as based on Green's identity, but they can also be interpreted as a weighted residual formulation as shown in this book. They are more versatile and general than indirect methods, which may be presented as a special case of direct formulations. To understand what they are and how they relate to other methods, let us consider the following potential problems,

$$\nabla^2 u = 0, \text{ in } \Omega \tag{1.1}$$

with boundary conditions,

$$u = \bar{u}, \text{ on } \Gamma_1$$

$$q = \frac{\partial u}{\partial n} = \bar{q}, \text{ on } \Gamma_2 \tag{1.2}$$

where the total boundary is $\Gamma = \Gamma_1 + \Gamma_2$ and Ω is the domain. The bar indicates known boundary condition values.

Consider now a weighting function w which we assume to be different from zero on Γ_1 and Γ_2. The function w is composed of a set of functions each of which is multiplied by an arbitrary parameter and added together. u indicates the function to be replaced by the approximate function, which is the result of adding together functions of a given set, each of them multiplied by an unknown parameter. The weighted residual expression corresponding to this problem can be shown to be (see Chapter 2)

$$\int_{\Omega} (\nabla^2 u) w \, d\Omega = \int_{\Gamma_2} (q - \overline{q}) w \, d\Gamma + \int_{\Gamma_1} (\overline{u} - u) \frac{\partial w}{\partial n} \, d\Gamma \tag{1.3}$$

Usually (but not always) the boundary conditions $u = \overline{u}$ are identically satisfied in methods such as finite elements, and the weighting function w is associated with the 'virtual' increments of u. Hence Equation (1.3) becomes

$$\int_{\Omega} (\nabla^2 u) \delta u \, d\Omega = \int_{\Gamma_2} (q - \overline{q}) \delta u \, d\Gamma \tag{1.4}$$

It is also usual to integrate by parts the left hand side of Equation (1.4) in order to reduce the order of differentiation. This gives

$$\int_{\Omega} \frac{\partial u}{\partial x_k} \frac{\partial \delta u}{\partial x_k} \, d\Omega = \int_{\Gamma_2} \overline{q} \delta u \, d\Gamma \tag{1.5}$$

Equation (1.5) is the starting expression for nearly all finite element models for potential problems.

Finite differences, on the other hand, can be interpreted as a special case of Equation (1.3) for which the weighting function w is written in terms of a series of Dirac functions and the derivatives of u are approximated in the usual finite difference way. If all boundary conditions are satisfied we can write (1.3) as,

$$\int_{\Omega} (\nabla^2 u) w \, d\Omega = 0 \tag{1.6}$$

or in terms of the Dirac function,

$$\int_{\Omega} (\nabla^2 u) \Delta_i \, d\Omega = 0 \tag{1.7}$$

where Δ_i represents the delta Dirac function at the 'i' point in the finite difference grid.

We can generalize finite differences by assuming that only the $u = \overline{u}$ conditions on Γ_1 are identically satisfied and integrate by parts the first terms of Equation (1.3). This gives,

$$\int_{\Omega} \frac{\partial u}{\partial x_k} \frac{\partial w}{\partial x_k} \, d\Omega = \int_{\Gamma_2} \overline{q} w \, d\Gamma \tag{1.8}$$

Finite difference approximations can now be used to represent u and w, which can now be interpreted as a 'virtual' increment or δu in much the same way as for finite elements. This technique is sometimes called the energy finite difference method because it is usually based on energy functionals, although these functionals are not explicitly needed.

Let us now integrate by parts Equation (1.3) without making any suppositions regarding the behaviour of the function w (except its differentiability). This gives,

$$-\int \frac{\partial u}{\partial x_k} \frac{\partial w}{\partial x_k} \, d\Omega = -\int_{\Gamma_2} \bar{q} w \, d\Gamma - \int_{\Gamma_1} q w \, d\Gamma + \int_{\Gamma_1} (\bar{u} - u) \frac{\partial w}{\partial n} \, d\Gamma \tag{1.9}$$

Integrating again by parts we obtain

$$\int_{\Omega} (\nabla^2 w) u \, d\Omega = -\int_{\Gamma_2} \bar{q} w \, d\Gamma - \int_{\Gamma_1} q w \, d\Gamma + \int_{\Gamma_2} u \frac{\partial w}{\partial n} \, d\Gamma + \int_{\Gamma_1} \bar{u} \frac{\partial w}{\partial n} \, d\Gamma \tag{1.10}$$

In other words we have now obtained the 'inverse' problem; the operator ∇^2 is now acting on w instead of on u. As w represents a set of weighting functions that we can choose it is possible to select them such that

$$\nabla^2 w \equiv 0 \tag{1.11}$$

Hence Equation (1.10) becomes,

$$\int_{\Gamma_2} \bar{q} w \, d\Gamma + \int_{\Gamma_1} q w \, d\Gamma = \int_{\Gamma_1} \bar{u} \frac{\partial w}{\partial n} \, d\Gamma + \int_{\Gamma_2} u \frac{\partial w}{\partial n} \, d\Gamma \tag{1.12}$$

Note that Equation (1.12) only applies on the boundary.

The function w can be chosen in many different ways but it is usual to take it as the fundamental solution. In this case another term (due to the concentrated load) will generally appear producing a Somigliani type identity. This relationship will be studied in detail in the following chapters but it is important to point out that the fundamental solution is not the only solution that satisfies Equation (1.11). Indeed in many cases it may be far more convenient to use simple functions that satisfy the governing equation rather than to apply the fundamental solution. This may be the case in layered media or other nonhomogeneous problems as well as for problems with complex governing equations.

These and other ideas will be explored in more detail in the following chapters but it is important to realise the many possibilities that the boundary methods offer. It is also important to be aware of the possibility of combining boundary with domain elements and how their relationship can be easily understood using weighted residual techniques.

2

Weighted residual methods

2.1 BASIC CONCEPTS

An operator is a process which when applied to some function u produces another function, say p. We can write,

$$\mathscr{L}(u) = p, \qquad \text{in } \Omega \tag{2.1}$$

where $\mathscr{L}(\ )$ is the operator and Ω represents the spatial coordinates x, y, z. We will usually consider differential operators although they can also be of integral type. An operator is linear if

$$\mathscr{L}(\alpha u_1 + \beta u_2) = \alpha\,\mathscr{L}(u_1) + \beta\,\mathscr{L}(u_2) \tag{2.2}$$

where α and β are numbers.

Consider now the problem represented by a set of homogeneous equations in the interior of a domain, Ω

$$\mathscr{L}(u) = 0, \qquad \text{in } \Omega \tag{2.3}$$

and define the inner product of this $\mathscr{L}(u)$ with another function v as,

$$\int \mathscr{L}(u) v \, d\Omega \tag{2.4}$$

Sometimes this is expressed as $\langle \mathscr{L}(u), v \rangle$. Other definitions of inner products are equally possible but this is the one we will use throughout the book. We can integrate by parts Equation (2.4) until the derivatives of u are eliminated. This leads to a 'transposed' form of the inner product and also to boundary terms. We write the result as,

$$\int \mathscr{L}(u) v \, d\Omega = \int u\, \mathscr{L}^*(v) \, d\Omega + \int_\Gamma [G(v)S(u) - G(u)S^*(v)]\, d\Gamma \tag{2.5}$$

where Γ is the exterior surface and S and G are differential operators due to the integration by parts. By definition $G(v)$ contains the v terms resulting from the first phase of the partial integration and $S(u)$ contains the corresponding u terms. Some examples are included below.

The operator $\mathscr{L}^*$ is called the *adjoint* of $\mathscr{L}$. If $\mathscr{L}^* = \mathscr{L}$, $\mathscr{L}$ is said to be *self-adjoint.* In this case $S = S^*$ also. Self-adjointness of an operator is analogous to symmetry of a matrix. In addition to determining whether the operator is self-adjoint, the partial integration also generates two different categories of boundary conditions. The set $G(u)$ prescribed are called the *essential* boundary conditions and $S(u)$ prescribed are the *non-essential* or natural boundary conditions; one can specify either type of boundary condition on the surface of a domain. However, the essential boundary conditions must be enforced at some point in order for the solution to be unique. Letting Γ_1 and Γ_2 represent complementary portions of the total surface, Γ, we can state the boundary conditions for the self-adjoint problem ($\mathscr{L}^* = \mathscr{L}$) as,

$$\left.\begin{array}{l} G(u) \text{ prescribed on } \Gamma_1 \\ S(u) \text{ prescribed on } \Gamma_2 \end{array}\right\} \Gamma_1 + \Gamma_2 = \Gamma \tag{2.6}$$

A self-adjoint operator is also *positive definite* if

$$\int (\mathscr{L}(u)) u \, d\Omega \geqslant 0 \tag{2.7}$$

for all u and only equal to zero for the trivial case $u \equiv 0$. To determine if $\mathscr{L}$ is positive definite we can integrate the inner product by parts until it contains derivatives of the same order. This operation is the midpoint in the transformation of $\mathscr{L}$ into $\mathscr{L}^*$. Positive definiteness is an extremeley valuable property in establishing solution schemes and also in constructing variational statements.

Example 2.1

Properties analogous to the self-adjointness and positive definiteness of operators can also be defined for matrices. These are that the matrices are symmetric and positive definite.

Consider a square matrix $\mathbf{A} = [a_{ij}]$. We say that $\mathbf{A}^\mathrm{T} = \mathbf{A}$, where $\mathbf{A}^\mathrm{T}$ (the transpose of A) is formed by interchanging the rows and columns of A. Symmetry requires $a_{ij} = a_{ji}$ but another way of defining symmetry is to require,

$$\langle \mathbf{Y}, \mathbf{A}\,\mathbf{X} \rangle \equiv \langle \mathbf{X}, \mathbf{A}\,\mathbf{Y} \rangle \tag{a}$$

where $\langle \, , \rangle$ is an inner product, the simplest of which is,

$$\mathbf{Y}^T \mathbf{A}\,\mathbf{X} = \mathbf{X}^T \mathbf{A}\,\mathbf{Y} \tag{b}$$

where **X**, **Y** are arbitrary vectors. We can transpose the right hand side of (b) which gives

$$\mathbf{Y}^T \mathbf{A} \mathbf{X} = \mathbf{Y}^T \mathbf{A}^T \mathbf{X} \tag{c}$$

This shows that Equation (a) is equivalent to $\mathbf{A}^T = \mathbf{A}$.

The positive definite character of the matrix can be investigated by writing

$$\mathbf{X}^T \mathbf{A} \mathbf{X} \geqslant 0 \tag{d}$$

This has to be valid for all **X** and equals 0 only when **X** is a null vector.

Example 2.2

Show that the operator $\mathscr{L}(\;) = -\mathrm{d}^2(\;)/\mathrm{d}x^2$ is self-adjoint and positive definite over the interval $0 < x < 1$.

Forming the inner product and integrating yields

$$\int_0^1 \mathscr{L}(u) v \,\mathrm{d}x = \int_0^1 \left(-\frac{\mathrm{d}^2 u}{\mathrm{d}x^2} \right) v \,\mathrm{d}x = -\left| \frac{\mathrm{d}u}{\mathrm{d}x} v \right|_0^1 + \int_0^1 \frac{\mathrm{d}u}{\mathrm{d}x} \frac{\mathrm{d}v}{\mathrm{d}x} \mathrm{d}x =$$

$$= \left| u \frac{\mathrm{d}v}{\mathrm{d}x} \right|_0^1 - \left| \frac{\mathrm{d}u}{\mathrm{d}x} v \right|_0^1 + \int_0^1 \left(-\frac{\mathrm{d}^2 v}{\mathrm{d}x^2} \right) u \,\mathrm{d}x \tag{a}$$

which shows that the operator is self-adjoint ($\mathscr{L} = \mathscr{L}^*$). Note that

$$G(u) = u, \qquad G(v) = v$$

$$S(u) = -\frac{\mathrm{d}u}{\mathrm{d}x}, \qquad S^*(v) = -\frac{\mathrm{d}v}{\mathrm{d}x} \tag{b}$$

$$\mathscr{L}(\;) = \mathscr{L}^*(\;) \quad \text{and } S(\;) = S^*(\;)$$

The essential boundary condition is u prescribed and the natural boundary condition is $-\mathrm{d}u/\mathrm{d}x$ prescribed.

If we take $u = v$ and homogeneous boundary conditions, the first integration in Equation (a) gives

$$\int_0^1 \mathcal{L}(u)u\mathrm{d}x = \int_0^1 \left(\frac{\mathrm{d}u}{\mathrm{d}x}\right)^2 \mathrm{d}x \tag{c}$$

which proves that $\mathcal{L}$ ( ) is positive definite.

Example 2.3

Let us now investigate the properties of the operator $\mathcal{L}(\ ) = \mathrm{d}^4(\ )/\mathrm{d}x^4$. We form the inner product

$$\int_0^1 \mathcal{L}(u)v\mathrm{d}x = \int_0^1 \frac{\mathrm{d}^4 u}{\mathrm{d}x^4} v\mathrm{d}x \tag{a}$$

Integrating by parts twice,

$$\int_0^1 \mathcal{L}(u)v\mathrm{d}x = \left| v\frac{\mathrm{d}^3 u}{\mathrm{d}x^3}\right|_0^1 - \left|\frac{\mathrm{d}v}{\mathrm{d}x}\frac{\mathrm{d}^2 u}{\mathrm{d}x^2}\right|_0^1 + \int_0^1 \frac{\mathrm{d}^2 v}{\mathrm{d}x^2}\frac{\mathrm{d}^2 u}{\mathrm{d}x^2}\mathrm{d}x \tag{b}$$

Integrating again by parts twice, we obtain

$$\int_0^1 \mathcal{L}(u)v\mathrm{d}x = \left| v\frac{\mathrm{d}^3 u}{\mathrm{d}x^3}\right|_0^1 - \left|\frac{\mathrm{d}v}{\mathrm{d}x}\frac{\mathrm{d}^2 u}{\mathrm{d}x^2}\right|_0^1 +$$

$$+ \left|\frac{\mathrm{d}^2 v}{\mathrm{d}x^2}\frac{\mathrm{d}u}{\mathrm{d}x}\right|_0^1 - \left|\frac{\mathrm{d}^3 v}{\mathrm{d}x^3}u\right|_0^1 + \int_0^1 \frac{\mathrm{d}^4 v}{\mathrm{d}x^4}u\mathrm{d}x \tag{c}$$

Note that for this case we have,

$$\int_0^1 \mathcal{L}(u)v\mathrm{d}x = \int_0^1 \mathcal{L}^*(v)u\mathrm{d}x + [G_1(v)S_1(u) - G_2(v)S_2(u)]_0^1 +$$

$$+ [S_2(v)G_2(u) - S_1(v)G_1(u)]_0^1 \tag{d}$$

Hence the essential boundary conditions are,

$$G_1(u) = u, \qquad G_2(u) = \frac{du}{dx}, \text{ prescribed} \tag{e}$$

and the natural boundary conditions are,

$$S_1(u) = \frac{d^3u}{dx^3}, \qquad S_2(u) = \frac{d^2u}{dx^2}, \text{ prescribed} \tag{f}$$

The operator $\mathscr{L}^*$ is

$$\mathscr{L}^*(\,) = \mathscr{L}(\,) = \frac{d^4(\,)}{dx^4} \tag{g}$$

If we now put $v = u$ in Equation (b) we obtain

$$\int_0^1 \mathscr{L}(u)u\,dx = \left| u \frac{d^3u}{dx^3} \right|_0^1 - \left| \frac{du}{dx} \frac{d^2u}{dx^2} \right|_0^1 + \int_0^1 \left(\frac{d^2u}{dx^2} \right)^2 dx \tag{h}$$

For u satisfying the homogeneous boundary conditions this integral is essentially positive, i.e.

$$\int_0^1 \mathscr{L}(u)u\,dx = \int_0^1 \left(\frac{d^2u}{dx^2} \right)^2 dx \geqslant 0 \tag{i}$$

The integral is only zero when $u = \alpha x + \beta$, hence for it to be >0 for all u we must restrain at least $u(0)$ and $u(l)$ or one of them and a (du/dx). This in fact means that all rigid body type movements have been suppressed.

2.2 WEIGHTED RESIDUAL METHODS

Weighted residual methods are numerical procedures for approximating the solution of a set of differential equations of the form

$$\mathscr{L}(u_0) = p, \qquad \text{in } \Omega \tag{2.8}$$

with boundary conditions

$$\text{essential} \quad G(u_0) = g \quad (\text{on } \Gamma_1)$$

$$\text{natural} \quad S(u_0) = q \quad (\text{on } \Gamma_2) \tag{2.9}$$

$\Gamma = \Gamma_1 + \Gamma_2$ is the external surface of the domain Ω and u_0 the exact solution. The function u_0 is first approximated by a set of functions $\phi_k(x)$, such that

$$u = \sum_{k=1}^{n} \alpha_k \phi_k \tag{2.10}$$

where α_k are undetermined parameters and ϕ_k are *linearly independent functions taken from a complete sequence* of functions such as

$$\phi_1(x), \phi_2(x), \ldots, \phi_n(x) \tag{2.11}$$

These functions are usually chosen so as to satisfy certain given conditions, called admissibility conditions, relating to the boundary conditions and the degree of continuity. The functions we consider belong to a *linear* space, i.e. we can combine them linearly,

$$\phi = \alpha\phi_1 + \beta\phi_2 \tag{2.12}$$

The inner product of two functions will be defined as

$$\int \phi_1(x)\phi_2(x)\mathrm{d}x \tag{2.13}$$

and a measure (norm) of the function ϕ can be taken as the square root of the inner product of ϕ function by itself and is denoted by $\|\phi\|$, i.e.

$$\|\phi\| = \sqrt{\left(\int \phi^2 \mathrm{d}x\right)} \tag{2.14}$$

A sequence of functions such as Equation (2.11) is said to be *linearly independent* if

$$\alpha_1\phi_1 + \alpha_2\phi_2 + \ldots + \alpha_n\phi_n = 0 \tag{2.15}$$

is true only when all α_i are zero.

A sequence of linearly independent functions is said to be *complete* if a number N and a set of constants α_i can be found such that, given an admissible but otherwise arbitrary function u_0, we have

$$\left\| u_0 - \sum_{i=1}^{N} \alpha_i \phi_i \right\| < \beta \tag{2.16}$$

where β is a small positive quantity.

Returning to Equation (2.8) we will initially require that these functions ϕ satisfy all the boundary conditions of the problem and have the necessary degree of continuity as to make the left hand side of Equation (2.8) different from zero. A procedure for relaxing the boundary conditions and continuity requirements will be discussed in the next section.

Substitution of Equation (2.10) into (2.8) produces an error function ϵ, which is called the *residual*, i.e.

$$\epsilon = \mathscr{L}(u) - p \neq 0 \tag{2.17}$$

Note that ϵ is equal to zero for the exact solution but not for the approximation. This ϵ error or residual is forced to be zero, in the average sense, by setting weighted integrals of the residual equal to zero, i.e.

$$\int \epsilon \psi_i \, dx = 0 \qquad i = 1, 2, \ldots, n \tag{2.18}$$

where ψ_i is a set of weighting functions, which are part of a linearly independent set. The solution will converge towards the exact solution as the number of terms increases.

Example 2.4

It is important to notice that an infinite set of orthogonal functions for instance is not necessarily complete, i.e. we can take an infinite number of functions but the solution will not always converge to the exact solution. To show this consider the following set

$$\phi_k = \sin \frac{k\pi x}{l} \qquad (k = 1, 2, \ldots, n) \tag{a}$$

If we assume that the solution is

$$u = \Sigma \alpha_k \phi_k \tag{b}$$

even when $k \to \infty$ we will not be able to reproduce the trivial solution $u =$ constant. If the possibility of u being a constant exists we need to take instead of Equation (a) the following function,

$$u = \alpha_0 \cdot 1 + \Sigma \alpha_k \phi_k \tag{c}$$

which now is complete. This example shows the difficulty of establishing the completeness of a given set of functions.

In what follows we will describe the more important approximation methods which are based on this idea of orthogonalization and are generically called weighted residual methods. The weighting function in each of them is chosen in a different way.

We will only consider symmetric, positive definite operators, for simplicity, but the methods can be equally well applied to other more general types of operators.

Method of moments

As was already mentioned the weighting function ψ_i in Equation (2.18) can be different from the approximating functions ϕ_i and any complete and linearly independent set of functions can be used. The simplest choice is the series $1, x, x^2, x^3 \ldots$ for a one dimensional problem. In this way successive higher 'moments' of the residual are forced to be zero.

$$\int (\mathscr{L}(u) - p)\psi_j \, dx = 0 \tag{2.19}$$

$$u = \Sigma \alpha_i \phi_i$$

$$\psi_j = x^j, \qquad j = 0, 1, 2, \ldots$$

The above technique is called the 'method of moments'. If the weighting series is any other set of functions, the method is simply called a weighted residual method.

Example 2.5

Consider the following second order equation, which applies in the domain $0 < x < 1$.

$$\mathscr{L}(u) - p = \frac{d^2u}{dx^2} + u + x = 0 \tag{a}$$

with boundary conditions,

$$u = 0 \text{ at } x = 0$$

$$u = 0 \text{ at } x = 1 \tag{b}$$

We first propose as an approximating function,

$$u = x(1 - x)(\alpha_1 + \alpha_2 x + \ldots) \tag{c}$$

which satisfies the boundary conditions for arbitrary α_i.

If only two terms of the approximation are taken,

$$u = x(1 - x)(\alpha_1 + \alpha_2 x) \tag{d}$$

The error or residual function is

$$\epsilon = \mathscr{L}(u) - p = x + (-2 + x - x^2)\alpha_1 + \\ + (2 - 6x + x^2 - x^3)\alpha_2 \tag{e}$$

We orthogonalize it with respect to 1 and x.

$$\int_0^1 \epsilon \cdot 1 \, dx = 0, \qquad \int_0^1 \epsilon \cdot x \, dx = 0 \tag{f}$$

After integration we find,

$$\begin{bmatrix} \frac{11}{6} & \frac{11}{12} \\ \frac{11}{12} & \frac{19}{20} \end{bmatrix} \begin{Bmatrix} \alpha_1 \\ \alpha_2 \end{Bmatrix} = \begin{Bmatrix} \frac{1}{2} \\ \frac{1}{3} \end{Bmatrix} \tag{g}$$

The solution of this system gives,

$$\alpha_1 = \frac{122}{649}, \qquad \alpha_2 = \frac{110}{649} \tag{h}$$

The approximate solution then becomes,

$$u = x(1 - x)\left\{\frac{122}{649} + \frac{110}{649}x\right\} \tag{i}$$

In Table 1.1 this solution has been compared against the exact one, i.e.

$$u_{\text{exact}} = \frac{\sin x}{\sin 1} - x \tag{j}$$

Table 2.1

x	*Approximate solution*	*Exact solution*
0.25	0.043191	0.044014
0.50	0.068181	0.069747
0.75	0.059084	0.060056

The collocation method

Instead of trying to satisfy the differential equations in an 'average' form, we can try to satisfy them at only a series of chosen points. These points are usually but not necessarily, evenly distributed in the domain.

Consider the approximating function

$$u = \Sigma\, \alpha_k \phi_k \tag{2.20}$$

where ϕ_k satisfy the boundary conditions.

Let us determine the values of α_i by enforcing the condition that

$$\epsilon = \mathscr{L}(u) - p = 0 \tag{2.21}$$

at n points in the domain. Note that in principle the number of α_i constants has to be the same as the number of collocation points chosen.

We can also express this condition by defining a Dirac, $\Delta(x_i)$, function such that $\Delta(x_i) = 0$ if $x \neq x_i$, and

$$\int_{x_i-c}^{x_i+c} \Delta(x_i)\mathrm{d}x = 1, \; c \to 0 \tag{2.22}$$

Thus we write the point collocation method as a weighted residual technique,

$$\int \epsilon \Delta_i \mathrm{d}x = 0 \qquad i = 1, 2, \ldots, n \tag{2.23}$$

For two dimensional problems $\Delta = \Delta(x_i, y_i)$ is a function of x and y.

Example 2.6

Let us solve the following Poisson's equation

$$\frac{\partial^2 u}{\partial x^2} + \frac{\partial^2 u}{\partial y^2} = p \tag{a}$$

with boundary conditions $u = 0$ at $x = \pm a$, $y = \pm b$.

We take the following approximation,

$$u = (x^2 - a^2)(y^2 - b^2)(\alpha_1 + \alpha_2(x^2 + y^2) + \ldots) \tag{b}$$

Consider first one term and a square region, $a = b$, for simplicity,

$$u^{(1)} = \alpha_1(x^2 - a^2)(y^2 - a^2) \tag{c}$$

$$\epsilon^{(1)} = \mathscr{L}(u) - p = 2\{(y^2 - a^2) + (x^2 - a^2)\}\alpha_1 - p \tag{d}$$

If we set $\epsilon^{(1)} = 0$ at $x = y = 0$, we obtain

$$\alpha_1 = -\frac{p}{4a^2} \tag{e}$$

For the second approximation we take,

$$u^{(2)} = (x^2 - a^2)(y^2 - a^2)\{\alpha_1 + \alpha_2(x^2 + y^2)\} \tag{f}$$

Thus,

$$\epsilon^{(2)} = \mathscr{L}(u) - p = -p + 2\alpha_1(x^2 + y^2 - 2a^2) +$$

$$+ \alpha_2\{6x^2y^2 + 2(x^4 + y^4) - 12a^2(x^2 + y^2) + 4a^4\} \tag{g}$$

Setting $\epsilon^{(2)} = 0$ at $x = y = 0$, and $x = y = a/2$, we have,

$$p = -4a^2\alpha_1 + 4a^4\alpha_2$$

$$p = -a^2 3\alpha_1 - \frac{3}{4}a^4\alpha_2 \tag{h}$$

$$\alpha_1 = -\frac{19}{60}\frac{p}{a^2}, \quad \alpha_2 = -\frac{1}{15}\frac{p}{a^4} \tag{i}$$

Thus the value of u at $x = y = 0$ for the first approximation is

$$u^{(1)} = -\frac{a^2 p}{4} = -0.250a^2 p \quad \text{(j)}$$

and for the second approximation

$$u^{(2)} = -\frac{19}{60}a^2 p = -0.317a^2 p \quad \text{(k)}$$

The exact value (see Example 2.8) is

$$u_{\text{exact}} = -\frac{36.64}{\pi^4}a^2 p \quad \text{(l)}$$

The method of sub-regions

This method is similar to the collocation method described above, but now instead of requiring that the error function be zero at certain points, we seek that the integral of the error function over different regions shall be zero. Thus we take

$$\int_{\Omega_i} \epsilon \mathrm{d}\Omega = 0 \qquad \text{for 'i' different regions } \Omega_i$$

$$\int_{\Omega_i} (\mathscr{L}(u) - p)\mathrm{d}\Omega = 0 \quad (2.24)$$

Example 2.7

Solve the equation

$$\frac{\mathrm{d}^2 u}{\mathrm{d}x^2} + u + x = 0 \quad \text{(a)}$$

with boundary conditions,

$$u(0) = u(1) = 0$$

Let

$$u^{(n)} = x(1 - x)(\alpha_1 + \alpha_2 x + \ldots) \quad \text{(b)}$$

For the first approximation we can take all the region $0 < x < 1$. Thus

$$\epsilon^{(1)} = (-2 + x - x^2)\alpha_1 + x \tag{c}$$

gives

$$\int_0^1 \{(-2 + x - x^2)\alpha_1 + x\}\, dx = -\frac{11}{6}\alpha_1 + \frac{1}{2} = 0 \tag{d}$$

$$\therefore \alpha_1 = \frac{3}{11}$$

and

$$u^{(1)} = \frac{3}{11}x(1 - x) \tag{e}$$

For the second approximation we can take two regions, $0 < x < ½$ and $0 < x < 1$. Thus we obtain,

$$\frac{1}{8} - \frac{11}{12}\alpha_1 + \frac{53}{192}\alpha_2 = 0$$

$$\frac{1}{2} - \frac{11}{6}\alpha_1 - \frac{11}{12}\alpha_2 = 0 \tag{f}$$

which gives,

$$\alpha_1 = \frac{97}{517}, \qquad \alpha_2 = \frac{24}{141} \tag{g}$$

and

$$u^{(2)} = x(1 - x)\left(\frac{291 + 264x}{1551}\right) \tag{h}$$

The values of $u^{(1)}$, $u^{(2)}$ are compared in Table 2.2.

Although at first sight the method of subregions would appear to be advantageous compared with collocation, in practice the advantage of averaging the error function to zero over set regions is offset by the fact

Table 2.2

x	$u^{(1)}$	$u^{(2)}$	u_{exact}
0.25	0.051	0.043	0.044016
0.50	0.068	0.068	0.069747
0.75	0.051	0.059	0.060056

that the error function is likely to change signs several times in the region and thus to cover up quite large errors.

The method of Galerkin

Galerkin's method is a particular weighted residual method for which the weighting functions are the same as the trial functions. Given the system of equations,

$$\mathscr{L}(u) - p = 0, \quad \text{in } \Omega \tag{2.25}$$

with natural and essential boundary conditions,

$$S(u) = q, \quad G(u) = g, \quad \text{in } \Gamma \tag{2.26}$$

The approximate function which satisfy the boundary conditions is,

$$u = \sum_{k=1}^{N} \alpha_k \phi_k \tag{2.27}$$

Substitution of Equation (2.27) into (2.25) will produce a residual ϵ.

$$\epsilon = \mathscr{L}(\Sigma \alpha_k \phi_k) - p \tag{2.28}$$

which is then orthogonalized with respect to the same approximating functions ϕ_i, i.e.

$$\int \{\mathscr{L}(\Sigma \alpha_k \phi_k) - p\} \phi_i \mathrm{d}\Omega = 0, \quad i = 1, 2 \ldots n \tag{2.29}$$

If $\mathscr{L}$ is a linear operator Equation (2.29) produces a system of linear equations from which the α_k coefficients can be obtained. For instance (2.29) becomes,

$$\int \alpha_k \{\mathscr{L}(\phi_k)\} \phi_i \mathrm{d}\Omega = \int p \phi_i \mathrm{d}\Omega \tag{2.30}$$

$$k = 1, 2 \ldots n \qquad i = 1, 2 \ldots n$$

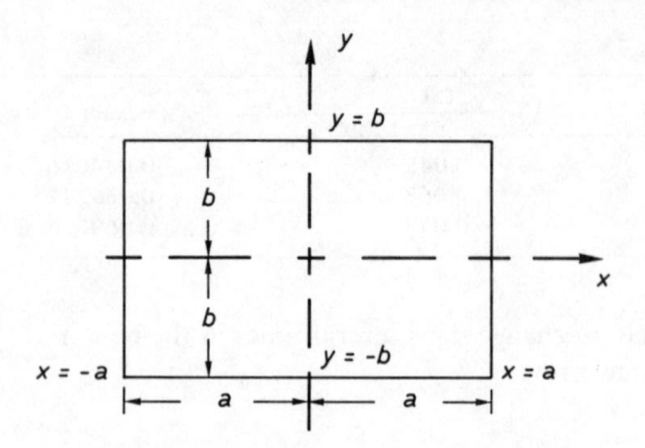

Fig.2.1. Two-dimensional domain for Poisson's equation

Contrary to other weighted residual methods in which the error is orthogonalized with respect to a set of functions different from the approximating functions, in Galerkin's procedure the weighting functions are the same as the approximating ones. This choice gives a physical significance to the Galerkin method in many engineering problems, as we shall see later.

Expression (2.29) which can be written as,

$$\int \{\mathscr{L}(u) - p\} \phi_i d\Omega = 0, \qquad i = 1, 2, \ldots, n \tag{2.31}$$

suggests the possibility of defining a series of *arbitrary* $\delta\alpha_i$ coefficients such that,

$$\delta u = \delta\alpha_1\phi_1 + \delta\alpha_2\phi_2 + \ldots + \delta\alpha_n\phi_n \tag{2.32}$$

This means that (2.31) can now be written in a more compact form, i.e.

$$\int \{\mathscr{L}(u) - p\} \delta u d\Omega = 0 \tag{2.33}$$

for arbitrary δu. It is understood that arbitrary δu is equivalent to $\delta\alpha_i\phi_i$ for $i = 1, 2 \ldots n$. We use this equivalent notation (i.e. (2.32)) with the Galerkin method.

Example 2.8

Consider the Poisson's equation, i.e.

$$\frac{\partial^2 u}{\partial x^2} + \frac{\partial^2 u}{\partial y^2} = p \tag{a}$$

with $u = 0$ at $x = \pm a$ and $y = \pm b$ (Figure 2.1).

As a first approximation one can take,

$$u = \alpha(x^2 - a^2)(y^2 - b^2) \tag{b}$$

The corresponding Galerkin statement can be written

$$\int_{-a}^{a}\int_{-b}^{b}\left\{\frac{\partial^2 u}{\partial x^2}+\frac{\partial^2 u}{\partial y^2}-p\right\}\delta u \mathrm{d}x\mathrm{d}y=\int_{-a}^{a}\int_{-b}^{b}\epsilon\delta u \mathrm{d}x\mathrm{d}y=0 \qquad \text{(c)}$$

where,

$$\delta u=\delta\alpha(x^2-a^2)(y^2-b^2) \qquad \text{(d)}$$

The error function is

$$\epsilon=2\alpha(y^2-b^2)+2\alpha(x^2-a^2)-p \qquad \text{(e)}$$

Hence Equation (c) becomes

$$\int_{-a}^{a}\int_{-b}^{b}\{2\alpha((y^2-b^2)+(x^2-a^2))-p\}(x^2-a^2)(y^2-b^2)\mathrm{d}x\mathrm{d}y=0 \qquad \text{(f)}$$

which gives after integration,

$$\frac{128}{45}\alpha\{a^3b^3(a^2+b^2)\}-\frac{16}{9}pa^3b^3=0 \qquad \text{(g)}$$

$$\alpha=\frac{5}{8}\frac{p}{(a^2+b^2)} \qquad \text{(h)}$$

Hence,

$$u=\frac{5}{8}\frac{p}{(a^2+b^2)}(x^2-a^2)(y^2-b^2) \qquad \text{(i)}$$

The value of u at the centre $(x = 0, y = 0)$ is

$$u_c=\frac{5}{8}p\frac{a^2b^2}{(a^2+b^2)} \qquad \text{(j)}$$

Let us now solve the same example using trigonometric series to approximate u. We assume, taking symmetry into consideration that

$$\alpha_{kl} \cos\left(k\frac{\pi}{2}\frac{x}{a}\right)\cos\left(l\frac{\pi}{2}\frac{y}{b}\right) \quad \text{(k)}$$

...sion satisfies the boundary conditions and the functions are ortho...al, i.e.

$$\int_{-a}^{a} \cos\left(m\frac{\pi}{2}\frac{x}{a}\right)\cos\left(n\frac{\pi}{2}\frac{x}{a}\right)\mathrm{d}x = 0, \text{ for } m \neq n$$

$$= a, \text{ for } m = n \quad \text{(l)}$$

Substituting Equation (k) into (c) we have that only the $\cos^2$ terms are going to be different from zero after integration. Hence,

$$\alpha_{kl}\int_{-a}^{a}\int_{-b}^{b}\left(\frac{k^2\pi^2}{4a^2}+\frac{l^2\pi^2}{4b^2}\right)\cos^2\left(\frac{k\pi}{2a}x\right)\cos^2\left(\frac{l\pi}{2b}y\right)\mathrm{d}x\mathrm{d}y -$$

$$-\int_{-a}^{a}\int_{-b}^{b} p\cos\left(\frac{k\pi}{2a}x\right)\cos\left(\frac{l\pi}{2b}y\right)\mathrm{d}x\mathrm{d}y = 0 \quad \text{(m)}$$

Integration gives,

$$\alpha_{kl} = \frac{64a^2b^2}{\pi^4 kl(k^2b^2+l^2a^2)}\,p(-1)^{\left(\frac{k+l}{2}\right)-1} \quad \text{(n)}$$

Note that the equations are uncoupled due to the orthogonality of the basis functions. The approximate solution is

$$u = \sum_k \sum_l \alpha_{kl}\cos\left(\frac{k\pi x}{2a}\right)\cos\left(\frac{l\pi y}{2b}\right) \quad \text{(o)}$$

When the number of terms tends to infinity we obtain the exact solution. For the centre point of a square domain ($a = b$) this value is

$$u_c = \left(32 - \frac{64}{15} + \frac{64}{162} + \ldots\right)\frac{a^2p}{\pi^4} \quad \text{(p)}$$

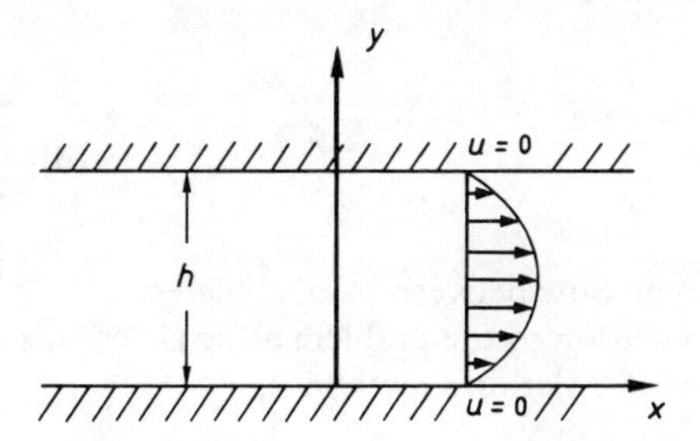

Fig.2.2. Flow in a channel

This result converges to

$$u_c = 28.70 \frac{a^2 p}{\pi^4} \tag{r}$$

which is the exact solution at the centre of the domain. Note that approximate solution (f) for a square domain can be written as,

$$u_c = 30.4 \frac{a^2 p}{\pi^4} \tag{s}$$

Example 2.9

Let us now consider the case of flow in a channel of unit depth when the velocity in the y direction is zero ($v = 0$, see Figure 2.2).

The continuity equation is now,

$$\frac{\partial u}{\partial x} + \frac{\partial v}{\partial y} = 0 \tag{a}$$

Thus for $v = 0$, $u = u(y)$ only.

For laminar confined flow and forced convection the momentum equation in the x direction can be expressed as,

$$\rho \left[u \frac{\partial u}{\partial x} + v \frac{\partial u}{\partial y} \right] = -\frac{\partial p}{\partial x} + \mu \frac{\partial^2 u}{\partial y^2} \tag{b}$$

From Equation (a) we have,

$$0 = -\frac{\partial p}{\partial x} + \mu \frac{\partial^2 u}{\partial y^2} \tag{c}$$

Integrating Equation (c) twice with respect to y and taking into account the boundary conditions we have,

$$u = \frac{1}{2}\,\frac{h^2}{\mu}\left(\frac{\partial p}{\partial x}\right)\left(\frac{y^2}{h^2} - \frac{y}{h}\right) \tag{d}$$

Equation (d) defines the Poiseuille type flow between parallel plates.

Let us now find an approximate solution to the problem using Galerkin's method. The weighted residual form of Galerkin's method is,

$$\int_0^h \left\{-\frac{\partial p}{\partial x} + \mu\,\frac{\partial^2 u}{\partial y^2}\right\}\delta u\,dy = 0 \tag{e}$$

We will now take the following approximate solution, which satisfies the boundary conditions,

$$u \simeq u_c \sin\left(\frac{\pi y}{h}\right) \tag{f}$$

Thus, expression (e) becomes,

$$\int_0^h \left[-\frac{\partial p}{\partial x}\sin\left\{\frac{\pi y}{h}\right\} - \mu\left\{\frac{\pi}{h}\right\}^2 u_c \sin^2\left\{\frac{\pi y}{h}\right\}\right]\delta u_c dy = 0 \tag{g}$$

$$-\frac{\partial p}{\partial x}\left\{\frac{2h}{\pi}\right\} - \mu\left\{\frac{\pi}{h}\right\}^2 u_c \frac{h}{2} = 0 \tag{h}$$

$$\therefore u_c = -\frac{4h^2}{\pi^3\mu}\,\frac{\partial p}{\partial x} \tag{i}$$

Hence we have the following velocity distribution,

$$u = -\frac{4h^2}{\pi^3\mu}\left\{\frac{\partial p}{\partial x}\right\}\sin\left\{\frac{\pi y}{h}\right\} \tag{j}$$

Velocity distributions (d) and (j) are compared in Table 2.3 for the case $(h^2/\mu)(\partial p/\partial x) = 1$.

Table 2.3. Poiseuille flow results

y	*Exact*	*Approximate*
0.0	0	0
0.1	-4.5×10^{-2}	-3.98648×10^{-2}
0.2	-8.0×10^{-2}	-7.58275×10^{-2}
0.3	−0.105	−0.10436
0.4	−0.12	−0.12269
0.5	−0.126	−0.12900

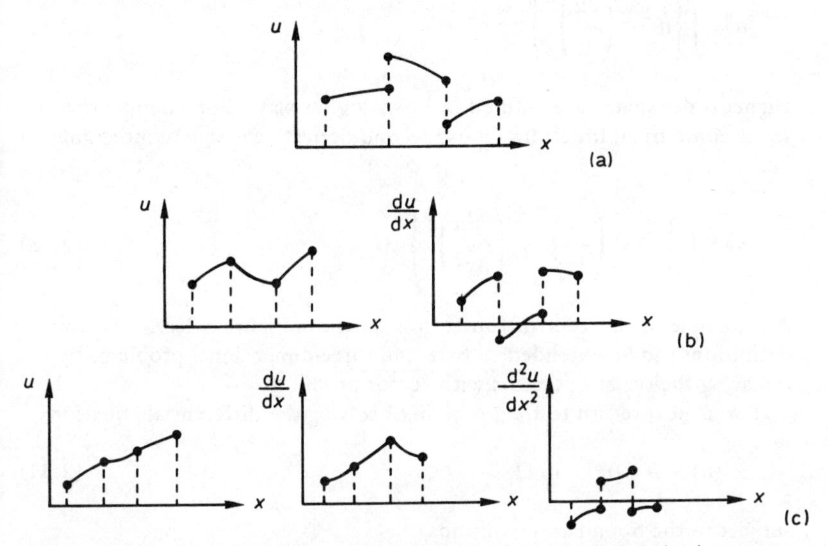

Fig.2.3. Types of function; (a) function square-integrable, (b) first derivative square-integrable, (c) second derivative square-integrable

2.3 WEAK FORMULATIONS

The examples treated in the previous section were restricted to self-adjoint operators and boundary conditions coinciding with the essential boundary conditions. Weighted residual methods are applicable for arbitrary operators and boundary conditions, and we will discuss in this section a general procedure for formulating weighted residual statements which allow only partial satisfaction of the boundary conditions and more significantly, the use of basis functions having relaxed requirements.

We need first to introduce a classification for the degree of continuity of a function. Consider a function $u(x)$ defined in a region Ω and having the shape illustrated in Figure 2.3(a). The function is discontinuous at discrete points but is finite throughout the region. Its norm satisfies the following condition,

$$\|u\| = \int_X u^2 \mathrm{d}x < \infty \tag{2.34}$$

All functions satisfying (2.34) are said to be square integrable.

Imposing restrictions on the continuity of the derivatives leads to a subset of spaces, called Sobolev spaces. The first of them contains all functions whose first derivative is square integrable (Figure 2.3(b)). Its definition is

$$\|u\| = \int \left\{ u^2 + \left(\frac{\mathrm{d}u}{\mathrm{d}x} \right)^2 \right\} \mathrm{d}x < \infty \tag{2.35}$$

Higher order spaces are defined in an analogous way. For example the next space contains all functions whose second derivative is square integrable, i.e.

$$\|u\| = \int \left\{ u^2 + \left(\frac{\mathrm{d}u}{\mathrm{d}x} \right)^2 + \left(\frac{\mathrm{d}^2 u}{\mathrm{d}x^2} \right)^2 \right\} \mathrm{d}x < \infty \tag{2.36}$$

An example of this type of function is shown in Figure 2.3(c). The above definitions can be extended to two- and three-dimensional problems by replacing the scalar operators with vector products.

Let us now return to the problem of solving the differential equation

$$\mathscr{L}(u) - p = 0, \qquad \text{in } \Omega \tag{2.37}$$

subject to the boundary conditions,

$$\text{natural} \qquad S(u) = q, \qquad \text{on } \Gamma_2 \tag{2.38}$$

$$\text{essential} \qquad G(u) = g, \qquad \text{on } \Gamma_1 \tag{2.39}$$

We will now assume that the trial functions for u satisfy the essential boundary conditions but not the natural conditions. At the same time we will release the continuity requirements of the function by lowering the order of the function space. We then obtain a *weak* solution. The lowering of the order of the function space is done by integrating by parts.

We can first write the following two error functions or residuals,

$$\begin{aligned} \epsilon &= \mathscr{L}(u) - p \text{ in the domain} \\ \epsilon_B &= S(u) - q \text{ on the boundary} \end{aligned} \tag{2.40}$$

Both residuals can now be weighted by a weighting function w.

$$\int \epsilon w \mathrm{d}\Omega = \int \epsilon_B w \mathrm{d}\Gamma \tag{2.41}$$

or

$$\int (\mathscr{L}(u) - p)w\mathrm{d}\Omega = \int (S(u) - q)w\mathrm{d}\Gamma \tag{2.42}$$

The function w belongs to the first space, Figure 2.3(a) and the type of space for u is determined by the highest derivatives of u.

In order to illustrate how the continuity requirements for u can be lowered, consider the following second order equation,

$$\mathscr{L}(u) - p = \frac{\mathrm{d}^2u}{\mathrm{d}x^2} + u + x = 0 \tag{2.43}$$

with natural boundary conditions,

$$\frac{\mathrm{d}u}{\mathrm{d}x} = q, \qquad \text{at } x = 1 \tag{2.44}$$

and essential boundary conditions

$$u = g, \qquad \text{at } x = 0 \tag{2.45}$$

We require the trial solution to satisfy the essential boundary condition. The test function satisfies the homogeneous form of the essential conditions.

$$\begin{aligned} u &= g, \qquad \text{at } x = 0 \\ w &= 0, \qquad \text{at } x = 0 \end{aligned} \tag{2.46}$$

The following statement can now be written,

$$\int \left(\frac{\mathrm{d}^2u}{\mathrm{d}x^2} + u + x\right) w\mathrm{d}x + \left| w\left(q - \frac{\mathrm{d}u}{\mathrm{d}x}\right)\right|_{x=1} = 0 \tag{2.47}$$

where w belongs to space (a) of Figure 2.3 and u to space (c). Integrating by parts one obtains,

$$\int \left\{(u + x)w - \frac{\mathrm{d}u}{\mathrm{d}x}\frac{\mathrm{d}w}{\mathrm{d}x}\right\} \mathrm{d}x + |qw|_{x=1} = 0 \tag{2.48}$$

where now u and w belong to space (b) of Figure 2.3, i.e. first derivative square integrable space. Integrating again we find,

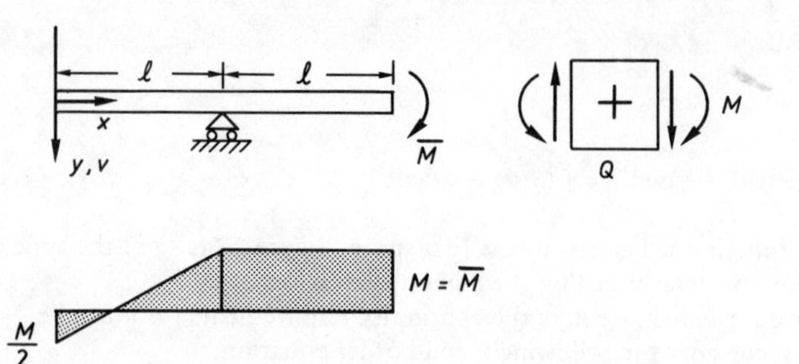

Fig.2.4. Beam subject to end-moment $\bar{M}$

$$\int\left\{(u+x)w+u\,\frac{\mathrm{d}^2w}{\mathrm{d}x^2}\right\}\mathrm{d}x+\left|qw-u\,\frac{\mathrm{d}w}{\mathrm{d}x}\right|_0^1=0 \tag{2.49}$$

where u now belongs to the function square integrable space and w to the second derivative square integrable one.

Scheme (2.48) is the most widely applied. Note that in Galerkin's method the trial and weighting functions are the same and one can replace w with δu.

Example 2.10

Consider the prismatic beam shown in Figure 2.4. The governing equation is

$$\mathscr{L}(v)=EI\,\frac{\mathrm{d}^4v}{\mathrm{d}x^4}=0 \tag{a}$$

where E is the modulus of elasticity and I the moment of inertia. The boundary conditions are,

$$v=\frac{\mathrm{d}v}{\mathrm{d}x}=0,\qquad \text{at } x=0$$

$$v=0,\qquad \text{at } x=l \tag{b}$$

$$M=EI\frac{\mathrm{d}^2v}{\mathrm{d}x^2}=\bar{M}\ \text{and}\ Q=-\frac{\mathrm{d}M}{\mathrm{d}x}=0,\qquad \text{at } x=2l.$$

Note that the last set of conditions are natural ones. Hence we will take approximate functions for v which satisfy the essential boundary conditions and have the necessary degree of continuity. We will start with the following weighted residual statement,

$$\int_0^{2l} \left\{ EI \frac{d^4 v}{dx^4} \right\} \delta v dx = \left[\left(EI \frac{d^3 v}{dx^3} + \bar{Q} \right) \delta v - \left(EI \frac{d^2 v}{dx^2} - \bar{M} \right) \frac{d\delta v}{dx} \right]_{x=2l} \tag{c}$$

Note that v here needs to be square integrable up to the fourth derivative. Integrating Equation (c) twice by parts we can reduce the order of the v function, obtaining

$$\int_0^{2l} EI \frac{d^2 v}{dx^2} \frac{d^2 \delta v}{dx^2} dx = \left[\bar{Q} \delta v + \bar{M} \frac{d\delta v}{dx} \right]_{x=2l} \tag{d}$$

This new expression allows us to approximate the natural boundary conditions and use second derivative square integrable function for our approximation.

The simplest choice for the v function is,

$$v^{(1)} = \alpha_1^{(1)} \phi_1 = \alpha_1^{(1)} (x^2 (x - l)) \tag{e}$$

Substituting this into Equation (d), taking into account that $\bar{Q} = 0$, will give

$$EI \int_0^{2l} \alpha_1^{(1)} (6x - 2l)^2 dx = 8l^2 \bar{M} \tag{f}$$

$$56 EI \alpha_1^{(1)} l^3 = 8l^2 \bar{M}$$

$$\therefore \alpha_1^{(1)} = \left(\frac{\bar{M}}{EIl} \right) \frac{1}{7} \tag{g}$$

The rotation and values of M and Q at $x = 2l$ can now be calculated. They are,

$$\left(\frac{dv}{dx} \right)_{x=2l} = 8\alpha_1^{(1)} l^2 = \left(\frac{\bar{M} l}{EI} \right) \frac{8}{7}$$

$$M = EI \left(\frac{d^2 v}{dx^2} \right)_{x=2l} = \frac{10}{7} \bar{M}, \qquad Q = -EI \left(\frac{d^3 v}{dx^3} \right)_{x=2l} = -6 \frac{\bar{M}}{7l} \tag{h}$$

We can see that the values of M and Q are quite different from the applied $\overline{M}$ and $\overline{Q}$ forces. The rotation however, is not too far off from the exact value.

$$\left(\frac{dv}{dx}\right)_{\substack{\text{exact}\\x=2l}} = 1.250\left(\frac{\overline{M}l}{EI}\right) \tag{i}$$

We can now improve upon these results by taking a second approximation for v, such that

$$v^{(2)} = \alpha_1^{(2)}\phi_1 + \alpha_2^{(2)}\phi_2 \tag{j}$$

where

$$\phi_1 = x^2(x - l), \qquad \phi_2 = x^3(x - l) \tag{k}$$

We can substitute these values into Equation (d) and obtain after integration two equations, corresponding to $\delta\alpha_1^{(2)}$ and $\delta\alpha_2^{(2)}$ increments. They are

$$\text{for } \delta\alpha_1^{(2)} \rightarrow 56\alpha_1^{(2)} + 152l\alpha_2^{(2)} = \frac{8\overline{M}}{EIl}$$

$$\text{for } \delta\alpha_2^{(2)} \rightarrow 152\alpha_1^{(1)} + 441.6l\alpha_2^{(2)} = \frac{20\overline{M}}{EIl} \tag{l}$$

Thus,

$$\alpha_1^{(2)} = 0.30315\left(\frac{\overline{M}}{EIl}\right), \qquad l\alpha_2^{(2)} = -0.05905\left(\frac{\overline{M}}{EIl}\right)$$

Finally the rotation and end forces at $x = 2l$ can be computed, i.e.

$$\left(\frac{dv}{dx}\right)_{x=2l} = 1.244\left(\frac{\overline{M}l}{EI}\right)$$

$$M = 0.9055\overline{M}, \qquad Q = 0.6614\frac{\overline{M}}{l} \tag{m}$$

The second trial solution gave excellent results for the displacements but convergence of the natural (forces) boundary conditions is not as rapid.

2.4 FUNCTIONALS AND RAYLEIGH–RITZ METHOD

Consider again the weighted residual statement corresponding to Galerkin's method

$$\int (\mathscr{L}(u) - p)\delta u \mathrm{d}\Omega = \int (S(u) - g)\delta u \mathrm{d}\Gamma \tag{2.50}$$

where the u functions are assumed to satisfy the essential boundary conditions and δu their homogeneous expression.

If $\mathscr{L}$ is a self-adjoint operator we will be able to write Equation (2.50) as the variation of a functional $F(u)$ to be defined.

In order to illustrate how this can be done consider the same equation as seen in (2.47), but with $w = \delta u$, i.e.

$$\int_0^1 \left(\frac{\mathrm{d}^2 u}{\mathrm{d}x^2} + u + x\right)\delta u \mathrm{d}x + \left[\left(q - \frac{\mathrm{d}u}{\mathrm{d}x}\right)\delta u\right]_{x=1} = 0 \tag{2.51}$$

Integrating by parts it gives,

$$\int_0^1 \left(-\frac{\mathrm{d}u}{\mathrm{d}x}\frac{\mathrm{d}\delta u}{\mathrm{d}x} + u\delta u\right)\mathrm{d}x + \int_0^l x\delta u \mathrm{d}x + |q\delta u|_{x=1} = 0 \tag{2.52}$$

Equation (2.52) can be written as the variation of a functional (or function of functions) F. We can first write,

$$\delta\left\{\frac{1}{2}\int_0^1 -\left(\frac{\mathrm{d}u}{\mathrm{d}x}\right)^2 \mathrm{d}x + \frac{1}{2}\int_0^1 (u^2)\mathrm{d}x + \int_0^1 xu\mathrm{d}x + |qu|_{x=1}\right\} = 0 \tag{2.53}$$

or

$$\delta F = 0$$

where the variation is carried out on the u function. The F functional is now

$$F(u) = \frac{1}{2}\int_0^1 \left\{-\left(\frac{\mathrm{d}u}{\mathrm{d}x}\right)^2 + u^2 + 2xu\right\}\mathrm{d}x + |qu|_{x=1} \tag{2.54}$$

It can also be written as

$$F = \int_0^l I\left(u, \frac{du}{dx}, x\right) dx + |qu|_{x=1} \tag{2.55}$$

where the integrand I is a function of u, du/dx and x.

If the functional F is known we can use the Rayleigh–Ritz method to obtain approximate solutions. The method consists in replacing the u variable in the F functional by

$$u = \sum_{k=1}^{n} \alpha_k \phi_k \tag{2.56}$$

and afterwards minimizing the functional with respect to the α_k variables.

The ϕ_k functions have to satisfy the essential boundary conditions of the problem and to be elements of a linearly independent complete set of functions.

For functional (2.55) the approximating function for u has to be square integrable up to the first derivative (function (b) of Figure 2.3).

Substituting the trial function and requiring $\delta F = 0$ means that,

$$\delta F = \frac{\partial F}{\partial \alpha_1} \delta\alpha_1 + \frac{\partial F}{\partial \alpha_2} \delta\alpha_2 + \ldots + \frac{\partial F}{\partial \alpha_n} \delta\alpha_n = 0 \tag{2.57}$$

or

$$\frac{\partial F}{\partial \alpha_i} = 0, \qquad \text{for } i = 1, 2, \ldots, n$$

In order to assess the convergence of the method, one has to take two or more trial functions. When the method is applied to a given functional with a minimum, one can measure convergence by comparing successive values of the functional obtained using the following sequence

$$\begin{aligned} u^{(1)} &= \alpha_1^{(1)} \phi_1 \\ u^{(2)} &= \alpha_1^{(2)} \phi_1 + \alpha_2^{(2)} \phi_2 \\ &\ldots \\ u^{(n)} &= \alpha_1^{(n)} \phi_1 + \alpha_2^{(n)} \phi_2 + \ldots + \alpha_n^{(n)} \phi_n \end{aligned} \tag{2.58}$$

where the n expansion includes all the functions contained in the previous expansions.

For positive definite, symmetrical functionals, we have, applying the above sequences,

$$F^{(1)} \geqslant F^{(2)} \geqslant \ldots \geqslant F^{(n)} \tag{2.59}$$

This behaviour is called monotonic, and the set of Equations (2.58) is called a minimizing sequence.

Example 2.11

Let us consider the problem of finding an approximate solution for the functional

$$F = \frac{1}{2}\int_0^1 \left\{ -\left(\frac{du}{dx}\right)^2 + u^2 + 2ux \right\} dx \tag{a}$$

with end conditions $u(0) = u(1) = 0$.

The governing equation corresponding to (a) can be obtained by operating on F using the variational notation, i.e.

$$\delta F = \int_0^1 \left\{ -\frac{du}{dx}\frac{d\delta u}{dx} + u\delta u + x\delta u \right\} dx = 0 \tag{b}$$

Integrating by parts,

$$\delta F = \int_0^1 \left\{ \frac{d^2u}{dx^2} + u + x \right\} \delta u dx = 0 \tag{c}$$

Thus,

$$\frac{d^2u}{dx^2} + u + x = 0 \tag{d}$$

is the governing equation corresponding to functional (a).

The simplest approximation is

$$u^{(1)} = \alpha_1^{(1)}(1 - x)x \tag{e}$$

Substituting Equation (e) into (a) gives

$$F = \frac{1}{2}\int_0^1 \{-(\alpha_1^{(1)})^2(2x-1)^2 + (\alpha_1^{(1)})^2(x^2-x)^2 - 2\alpha_1^{(1)}x(x^2-x)\}\,dx \quad \text{(f)}$$

After integration we obtain

$$F = -\frac{3}{10}(\alpha_1^{(1)})^2 + \frac{2}{12}\alpha_1^{(1)} \quad \text{(g)}$$

We can now take the variation of F, i.e.

$$\delta F = \frac{\partial F}{\partial \alpha_1^{(1)}}\,\delta\alpha_1^{(1)} = 0, \frac{\partial F}{\partial \alpha_1^{(1)}} = -\frac{6}{10}\alpha_1^{(1)} + \frac{2}{12} = 0 \quad \text{(h)}$$

and obtain

$$\alpha_1^{(1)} = \frac{5}{18} \quad \text{(i)}$$

Hence

$$u^{(1)} = \frac{5}{18}x(1-x) \quad \text{(j)}$$

As a second approximation we take,

$$u^{(2)} = x(1-x)(\alpha_1^{(2)} + \alpha_2^{(2)}x) \quad \text{(k)}$$

After substituting this expression into Equation (a), integrating and taking derivatives with respect to $\alpha_1^{(2)}$ and $\alpha_2^{(2)}$ we obtain

$$\frac{\partial F}{\partial \alpha_1^{(2)}} = 0 \rightarrow \frac{3}{10}\alpha_1^{(2)} + \frac{3}{20}\alpha_2^{(2)} = \frac{1}{12}$$

$$\frac{\partial F}{\partial \alpha_2^{(2)}} = 0 \rightarrow \frac{3}{20}\alpha_1^{(2)} + \frac{13}{105}\alpha_2^{(2)} = \frac{1}{20} \quad \text{(l)}$$

Table 2.4

$x =$	u_{exact}	$u^{(1)}$	$u^{(2)}$
0.25	0.044014	0.0521	0.0440
0.50	0.069747	0.0694	0.0698
0.75	0.060056	0.0521	0.0600

Thus

$$\alpha_1^{(2)} = \frac{71}{369}, \qquad \alpha_2^{(2)} = \frac{7}{41} \tag{m}$$

The approximate solution is

$$u^{(2)} = x(1 - x)\left[\frac{71}{369} + \frac{7}{41}x\right] \tag{n}$$

The exact solution for Equation (d) with boundary condition $u(0) = u(1) = 0$ is

$$u = \frac{\sin x}{\sin 1} - x \tag{o}$$

The comparison of this exact solution against the two approximations is shown in Table 2.4.

2.5 SUBSIDIARY CONDITIONS

In certain cases we want a functional or function to satisfy additional conditions, called subsidiary conditions and they can be introduced using Lagrange multipliers.

Let us review briefly what the Lagrange multipliers are before using them with functionals. Consider a function $f(x, y)$ from which we want to obtain the stationary value,

$$\mathrm{d}f = \frac{\partial f}{\partial x}\,\mathrm{d}x + \frac{\partial f}{\partial y}\,\mathrm{d}y \tag{2.60}$$

subject to the *subsidiary condition*

$$g(x, y) = 0 \tag{2.61}$$

We can differentiate (2.61) and obtain,

$$dg = \frac{\partial g}{\partial x}\,dx + \frac{\partial g}{\partial y}\,dy \tag{2.62}$$

Let us multiply Equation (2.62) by the parameter λ and then add them to Equation (2.60). Thus we have,

$$\left(\frac{\partial f}{\partial x} + \lambda\frac{\partial g}{\partial x}\right)dx + \left(\frac{\partial f}{\partial y} + \lambda\frac{\partial g}{\partial y}\right)dy = 0 \tag{2.63}$$

We have obtained three equations which together with (2.61) permit us to determine the unknowns, x, y, λ. The parameter λ is a Lagrangian multiplier and it usually has a physical meaning.

One can now write the stationary problem in another way. Let us define a new function

$$f + \lambda g \tag{2.64}$$

Differentiating (2.64) with respect to x, y, and λ we obtain the following three equations,

$$\frac{\partial f}{\partial x} + \lambda\frac{\partial g}{\partial x} = 0$$

$$\frac{\partial f}{\partial y} + \lambda\frac{\partial g}{\partial y} = 0$$

$$g = 0 \tag{2.65}$$

This gives a complete set of equations to be solved.

We can now extend the use of Lagrangian multipliers to the case of functionals. Assume for instance a functional

$$F = \int_{x_1}^{x_2} I\left(u, \frac{du}{dx}, x\right)dx \tag{2.66}$$

under the subsidiary condition

$$G = \int_{x_1}^{x_2} J\left(u, \frac{du}{dx}, x\right)dx \tag{2.67}$$

By using Lagrangian multipliers a new functional can be written, i.e.

$$F + \lambda G \tag{2.68}$$

Varying this new functional we have,

$$\delta(F + \lambda G) = 0 \tag{2.69}$$

or

$$\delta F + \lambda \delta G = 0$$

plus the condition $G = 0$.

Example 2.12

Consider a two dimensional solid for which the F functional is the total potential energy, i.e.

$$F = \frac{1}{2}\iint (\sigma_x \epsilon_x + \sigma_y \epsilon_y + \tau\gamma)\mathrm{d}\Omega - \int (\overline{p}_x u + \overline{p}_y v)\mathrm{d}\Gamma \tag{a}$$

where σ are the direct stresses, τ the shear stress, ϵ's are the extensional strains and γ the shear strains. $\overline{p}_x$ and $\overline{p}_y$ are applied forces on the boundary Γ and u and v are the displacements in x and y directions.

Assume that we want to apply to F the *incompressibility* condition,

$$\frac{\partial u}{\partial x} + \frac{\partial v}{\partial y} = 0 \tag{b}$$

Hence

$$G = \iint \left(\frac{\partial u}{\partial x} + \frac{\partial v}{\partial y}\right) \mathrm{d}\Omega = 0$$

The new functional is

$$\Pi = F + \lambda G = \frac{1}{2}\iint (\sigma_x \epsilon_x + \sigma_y \epsilon_y + \tau\gamma)\mathrm{d}\Omega - \int_{\Gamma_2} (\overline{p}_x u + \overline{p}_y v)\mathrm{d}\Gamma +$$

$$+ \lambda \iint \left(\frac{\partial u}{\partial x} + \frac{\partial y}{\partial y}\right) \mathrm{d}\Omega \tag{c}$$

The variation of Equation (c) gives

$$\delta\Pi = \iint (\sigma_x \delta\epsilon_x + \sigma_y \delta\epsilon_y + \tau\delta\gamma)\mathrm{d}\Omega - \int_{\Gamma_2} (\bar{p}_x \delta u + \bar{p}_y \delta v)\mathrm{d}\Gamma +$$

$$+ \iint \lambda \left(\frac{\partial \delta u}{\partial x} + \frac{\partial \delta v}{\partial y}\right) \mathrm{d}\Omega + \iint \delta\lambda \left(\frac{\partial u}{\partial x} + \frac{\partial v}{\partial y}\right) \mathrm{d}\Omega \tag{d}$$

Integrating by parts we obtain

$$\delta\Pi = -\iint \left\{\left(\frac{\partial \sigma_x}{\partial x} + \frac{\partial \tau}{\partial y}\right)\delta u + \left(\frac{\partial \sigma_y}{\partial y} + \frac{\partial \tau}{\partial x}\right)\delta v\right\} \mathrm{d}\Omega +$$

$$+ \int_{\Gamma_2} \{(p_x - \bar{p}_x)\delta u + (p_y - \bar{p}_y)\delta v\} \,\mathrm{d}\Gamma +$$

$$+ \iint \lambda \left(\frac{\partial \delta u}{\partial x} + \frac{\partial \delta v}{\partial y}\right) \mathrm{d}\Omega + \iint \delta\lambda \left(\frac{\partial u}{\partial x} + \frac{\partial v}{\partial y}\right) \mathrm{d}\Omega \tag{e}$$

The σ_x, σ_y stresses can be written as,

$$\sigma_x = \sigma_x' - p$$

$$\sigma_y = \sigma_y' - p \tag{f}$$

where σ' are the deviatoric components and p is the pressure, equal to the mean stresses with sign changed.

Thus Equation (e) after integrating by parts the pressure terms in the first term, becomes,

$$\delta\Pi = -\iint \left\{\left(\frac{\partial \sigma_x'}{\partial x} + \frac{\partial \tau}{\partial y}\right)\delta u + \left(\frac{\partial \sigma_y'}{\partial y} + \frac{\partial \tau}{\partial x}\right)\delta v\right\} \mathrm{d}\Omega +$$

$$+ \int_{\Gamma_2} \{(p_x' - \bar{p}_x)\delta u + (p_y' - \bar{p}_y)\delta v\} \,\mathrm{d}\Gamma +$$

$$+ \iint \delta\lambda \left(\frac{\partial u}{\partial x} + \frac{\partial y}{\partial y}\right) \mathrm{d}\Omega + \iint (\lambda - p) \left(\frac{\partial \delta u}{\partial x} + \frac{\partial \delta v}{\partial y}\right) \mathrm{d}\Omega \tag{g}$$

where p'_x p'_y are the deviatoric components of the surface forces ($p'_x = p_x + p \cos(n, x)$, $p'_y = p_y + p \cos(n, y)$). The Lagrangian λ can now be identified. It is

$$\lambda = p \tag{h}$$

Hence Equation (c) can be written as

$$\Pi = \frac{1}{2}\iint \{(\sigma'_x - p)\epsilon_x + (\sigma'_y - p)\epsilon_y + \tau\gamma\}\, d\Omega - \\ - \int_{\Gamma_2} (\overline{p}_x u + \overline{p}_y v) d\Gamma + \iint p\left(\frac{\partial u}{\partial x} + \frac{\partial v}{\partial y}\right) d\Omega \tag{i}$$

which is the functional we could use for an incompressible solid, such as some soils.

Example 2.13

Let us now consider a potential problem defined by the governing equation,

$$\frac{\partial^2 u}{\partial x^2} + \frac{\partial^2 u}{\partial y^2} = 0 \quad \text{in } \Omega \tag{a}$$

and the boundary condition,

$$q = \frac{\partial u}{\partial n} = \overline{q} \quad \text{(natural)} \quad \text{on } \Gamma_2$$

$$u = \overline{u} \quad \text{(essential)} \quad \text{on } \Gamma_1 \tag{b}$$

The weighted residual expression for the governing equations plus the natural boundary condition is,

$$\iint \left(\frac{\partial^2 u}{\partial x^2} + \frac{\partial^2 u}{\partial y^2}\right) w d\Omega = \int_{\Gamma_2} \left(\frac{\partial u}{\partial n} - \overline{q}\right) w d\Gamma \tag{c}$$

and the essential boundary conditions are assumed to be satisfied.

Let us now try to approximate the essential boundary conditions by including them in the expression (c). The condition is

$$u - \bar{u} = 0 \tag{d}$$

and the Lagrangian multiplier is λ.

The functional for Equation (c) is

$$\Pi = \frac{1}{2}\iint \left\{ \left(\frac{\partial u}{\partial x}\right)^2 + \left(\frac{\partial u}{\partial y}\right)^2 \right\} d\Omega - \int_{\Gamma_2} \bar{q} u \, d\Gamma \tag{e}$$

The new functional with the λ Lagrangian multiplier is

$$\Pi^* = \Pi + \lambda G =$$

$$= \frac{1}{2}\iint \left\{ \left(\frac{\partial u}{\partial x}\right)^2 + \left(\frac{\partial u}{\partial y}\right)^2 \right\} d\Omega - \int_{\Gamma_2} \bar{q} u \, d\Gamma + \int_{\Gamma_1} \lambda(u - \bar{u}) d\Gamma \tag{f}$$

Varying Equation (f) we obtain

$$\delta\Pi^* = \iint \left\{ \left(\frac{\partial u}{\partial x}\right)\left(\frac{\partial \delta u}{\partial x}\right) + \left(\frac{\partial u}{\partial y}\right)\left(\frac{\partial \delta u}{\partial y}\right) \right\} d\Omega - \int_{\Gamma_2} \bar{q} \delta u \, d\Gamma +$$

$$+ \int_{\Gamma_1} \delta\lambda(u - \bar{u}) d\Gamma + \int_{\Gamma_1} \lambda \delta u \, d\Gamma = 0 \tag{g}$$

Integrating the first term by parts, we find

$$\delta\Pi^* = -\iint \left(\frac{\partial^2 u}{\partial x^2} + \frac{\partial^2 u}{\partial y^2} \right) \delta u \, d\Omega - \int_{\Gamma_2} (\bar{q} - q) \delta u \, d\Gamma +$$

$$+ \int_{\Gamma_1} \delta\lambda(u - \bar{u}) d\Gamma + \int_{\Gamma_1} (\lambda + q) \delta u \, d\Gamma = 0 \tag{h}$$

The last integral gives

$$\lambda = -q = -\frac{\partial u}{\partial n} \tag{i}$$

as a condition. Hence we can write Equation (h) as

$$\delta\Pi^* = -\iint\left(\frac{\partial^2 u}{\partial x^2} + \frac{\partial^2 u}{\partial y^2}\right)\delta u \mathrm{d}\Omega - \int_{\Gamma_2} (\bar{q} - q)\delta u \mathrm{d}\Gamma -$$

$$- \int_{\Gamma_1} (u - \bar{u}) \frac{\partial \delta u}{\partial n} \mathrm{d}\Gamma = 0 \qquad \text{(j)}$$

If instead of δu we have a more general weighting function w the final expression would be,

$$\delta\Pi^* = -\iint\left(\frac{\partial^2 u}{\partial x^2} + \frac{\partial^2 u}{\partial y^2}\right) w \mathrm{d}\Omega - \int_{\Gamma_2} (\bar{q} - q) w \mathrm{d}\Gamma -$$

$$- \int_{\Gamma_1} (u - \bar{u}) \frac{\partial w}{\partial n} \mathrm{d}\Gamma = 0 \qquad \text{(k)}$$

This expression is the starting point for boundary element models.

2.6 BOUNDARY METHODS

Up to now we have considered functions which satisfy the boundary conditions and are approximate in the domain, that is do not satisfy exactly the governing equations. Conversely, we could propose functions which identically satisfy the governing equations and only approximately the boundary conditions.

Consider the governing equation studied in Example 2.13, i.e.

$$\frac{\partial^2 u}{\partial x^2} + \frac{\partial^2 u}{\partial y^2} = 0, \quad \text{in } \Omega \qquad (2.70)$$

with boundary conditions,

$$q = \frac{\partial u}{\partial n} = \bar{q} \quad \text{on } \Gamma_2$$

$$u = \bar{u} \quad \text{on } \Gamma_1 \qquad (2.71)$$

The weighted residual statement can be written as (see Example 2.13)

$$\iint \left(\frac{\partial^2 u}{\partial x^2} + \frac{\partial^2 u}{\partial y^2} \right) w \, d\Omega + \int_{\Gamma_2} (\bar{q} - q) w \, d\Gamma + \int_{\Gamma_1} (u - \bar{u}) \frac{\partial w}{\partial n} \, d\Gamma = 0 \tag{2.72}$$

Integrating the first integral in Equation (2.72) twice by parts we obtain,

$$\iint \left(\frac{\partial^2 w}{\partial x^2} + \frac{\partial^2 w}{\partial y^2} \right) u \, d\Omega = - \int_{\Gamma_2} \bar{q} w \, d\Gamma - \int_{\Gamma_1} \frac{\partial u}{\partial n} w \, d\Gamma + \int_{\Gamma_2} u \frac{\partial w}{\partial n} \, d\Gamma + \int_{\Gamma_1} \bar{u} \frac{\partial w}{\partial n} \, d\Gamma \tag{2.73}$$

We can now assume that the w functions identically satisfy the governing equation $(\partial^2 w/\partial x^2) + (\partial^2 w/\partial y^2) = 0$ but not the boundary conditions. This gives

$$\int_{\Gamma_2} \bar{q} w \, d\Gamma + \int_{\Gamma_1} \frac{\partial u}{\partial n} w \, d\Gamma = \int_{\Gamma_2} u \frac{\partial w}{\partial n} \, d\Gamma + \int_{\Gamma_1} \bar{u} \frac{\partial w}{\partial n} \, d\Gamma \tag{2.74}$$

The problem has now been reduced to a boundary problem and we can find the solution of the problem u.

If one takes $w = \delta u$ we have the Trefftz Method. Hence Equation (2.74) becomes,

$$\int_{\Gamma_2} \bar{q} \delta u \, d\Gamma + \int_{\Gamma_1} \frac{\partial u}{\partial n} \delta u \, d\Gamma = \int_{\Gamma_2} u \frac{\partial \delta u}{\partial n} \, d\Gamma + \int_{\Gamma_1} \bar{u} \frac{\partial \delta u}{\partial n} \, d\Gamma \tag{2.75}$$

Expression (2.74) is however, more general than the usual Trefftz formulation and will be used later on for some of the boundary elements approaches.

Note that Equations (2.74) or (2.75) can be written in a more compact way, i.e.

$$\int_{\Gamma} q w \, d\Gamma = \int_{\Gamma} u \frac{\partial w}{\partial n} \, d\Gamma \tag{2.76}$$

and

$$\int_{\Gamma} q \delta u \, d\Gamma = \int_{\Gamma} u \frac{\partial \delta u}{\partial n} \, d\Gamma \tag{2.77}$$

where the Γ boundary is $\Gamma_1 + \Gamma_2$. The q and u functions will become $\bar{q}$

and $\bar{u}$ on the Γ_2 and Γ_1 part of the boundary respectively.

Example 2.14

Consider the Poisson equation

$$\mathscr{L}(u) - p = \frac{\partial^2 u}{\partial x^2} + \frac{\partial^2 u}{\partial y^2} - p = 0 \qquad \text{(a)}$$

with boundary conditions $u = 0$ on Γ $(x = \pm 1, y = \pm 1)$.

Let us first reduce the problem to a Laplace equation problem. This can be done by defining a new function v, such that

$$u = \frac{p}{4}(x^2 + y^2) + v \qquad \text{(b)}$$

Thus Equation (a) becomes,

$$\frac{\partial^2 v}{\partial x^2} + \frac{\partial^2 v}{\partial y^2} = 0 \qquad \text{(c)}$$

with $v = -(p/4)(x^2 + y^2)$ on the boundaries $x = \pm 1$ and $y = \pm 1$.

We now approximate v by a trial function that satisfies the Laplace equation, for instance,

$$v = \alpha_1 \phi_1 \qquad \text{(d)}$$

with

$$\phi_1 = x^4 - 6x^2y^2 + y^4 \qquad \text{(e)}$$

We now have to satisfy the boundary integrals. Because of symmetry we can consider only $x = 1$, $0 < y < 1$. As all the boundary is of type Γ_1, Equation (2.75) reduces to,

$$\int_{\Gamma_1} \frac{\partial v}{\partial n} \delta v \, d\Gamma = \int_{\Gamma_1} \bar{v} \frac{\partial \delta v}{\partial n} \, d\Gamma \qquad \text{(f)}$$

or

$$\int_{\Gamma_1} \left(\frac{\partial v}{\partial n} \delta v - \bar{v} \frac{\partial \delta v}{\partial n} \right) d\Gamma = 0 \qquad \text{(g)}$$

For the boundary $x = 1$, we can write

$$\int_0^1 \left\{ \frac{\partial v}{\partial x}\,\delta v - \bar{v}\,\frac{\partial \delta v}{\partial x} \right\} \mathrm{d}y = 0 \tag{h}$$

Whence,

$$\int_0^1 \left\{ (4 - 12y^2)\alpha_1(1 - 6y^2 + y^4)\delta\alpha_1 + \frac{p}{4}(1 + y^2)\delta\alpha_1(4 - 12y^2) \right\} \mathrm{d}y = 0 \tag{i}$$

or

$$\int_0^1 \left\{ \frac{p}{4}(1 + y^2) + \alpha_1(1 - 6y^2 + y^4) \right\} (4 - 12y^2)\mathrm{d}y = 0 \tag{j}$$

Integrating we have,

$$-\frac{1}{5}\,p + \alpha_1\,\frac{144}{35} = 0 \tag{k}$$

$$\therefore \alpha_1 = \frac{7}{144}\,p \tag{l}$$

This gives,

$$v = \alpha_1\phi_1 = \frac{7}{144}\,p(x^4 - 6x^2y^2 + y^4) \tag{m}$$

The magnitudes of v and u may however be displaced from the correct solution by a constant magnitude. In order to determine this constant we can carry out the following integration on the $x = 1$ side.

$$\int_0^1 \{\bar{v} - v + c\}\mathrm{d}y = 0 \tag{n}$$

or

$$\int_0^1 \left\{ -\frac{p}{4}(1 + y^2) + \alpha_1(1 - 6y^2 + y^4) + c \right\} dy = 0 \tag{o}$$

After integration we find

$$c + \frac{p}{3} - 0.039\,p = 0, \; c = 0.039\,p - \frac{p}{3} \tag{p}$$

Note that the value of v is equal to c at $x = y = 0$.

BIBLIOGRAPHY

Biezeno, C. B. and J. J. Koch, 'Over eeen Nieuwe Methode ter Brerkening van Vlokke Platen met Toepassing op. Eukele voor de Technick Belangrijke Belastingsgevallen'. *Ing. Grav.* **38**, 25–36 (1923)

Finlayson, B. A., *The Method of Weighted Residuals and Variational Principles*, Academic Press (1972)

Trefftz, E., 'Ein Gegenstrück zum Ritzschen Verfahren'. *Proc. 2nd Int. Congress Appl. Mech.*, Zurich (1926)

Courant, R. and Hilbert, D., *Methods of Mathematical Physics*, Vol.1, Wiley, New York (1953)

Kantorovich, L. V. and Krylov, V. I., *Approximate Methods of Higher Analysis*, Noordhoff Ltd. (1958)

Connor, J. J. and C. A. Brebbia, *Finite Element Techniques for Fluid Flow*, 2nd ed. Newnes–Butterworths, London (1977)

3

Potential problems

3.1 INTRODUCTION

In the previous chapter we have presented the boundary solution technique as a weighted residual method and given an example in which a solution satisfying the governing equation, but otherwise arbitrary, was chosen as our approximate solution. We will now try to generalise this idea and work with a certain type of influence solution. The new solutions represent an applied charge concentrated at a point and is called the fundamental solution.

Numerous papers and other works have been published in recent years on this type of boundary methods[1–9]. These methods are presented under different names such as "boundary integral equation methods", "boundary integral solutions', etc. In its most general form this technique consists of subdividing the boundary of the region under consideration into a series of elements; hence the name "boundary element techniques" seems more appropriate and is the one used in this book. In all the references the method is presented as a method unrelated to other analytical techniques. However, with the background of Chapter 2 we can show that it is a particular application of weighted residual techniques. In this way its relationship with other methods of analysis, such as finite elements, becomes clear.

In this chapter only the boundary element formulation of potential problems will be discussed. The basic ideas are similar for any other engineering problem and the applications of the technique in elasticity problems will be discussed in Chapter 4.

Usually the technique produces a singularity when the equation using the known (fundamental) solution is particularized on the boundary. A method is presented in this chapter (Section 3.2) to integrate the fundamental solution on the boundary in a simpler way than the one previously shown. The same integration technique will be applied in elasticity problems (Section 4.2) and considerably simplifies the derivations.

This chapter presents and discusses some potential solutions using different types of boundary elements, i.e. constant and linear variation. The applications emphasize the simplicity of boundary element techniques and the way in which they can be applied in engineering. Some examples also illustrate the advantages of boundary elements over finite elements.

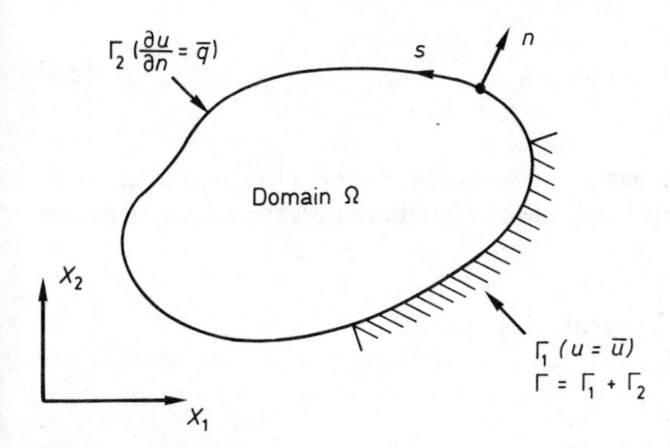

Fig.3.1. Definitions

3.2 BASIC RELATIONSHIPS

To illustrate how the boundary element formulation can be deduced as a weighted residual technique consider a potential function u on a domain Ω which is required to satisfy the following governing equation:

$$\nabla^2 u = 0, \quad \text{in } \Omega \tag{3.1}$$

The boundary conditions corresponding to this problem are of two types (Figure 3.1).

(a) Essential conditions, such as $u = \bar{u}$ on Γ_1

(b) Natural conditions of the type $\partial u/\partial n = \bar{q}$ on Γ_2

The total boundary is $\Gamma = \Gamma_1 + \Gamma_2$.

A weighting function u^* can now be introduced such that it has continuous first derivatives within Ω. Later this function will also be required to satisfy the governing equation. The following weighted residual statement can now be written,

$$\int_{\Omega} (\nabla^2 u) u^* \mathrm{d}\Omega = \int_{\Gamma_2} (q - \bar{q}) u^* \mathrm{d}\Gamma - \int_{\Gamma_1} (u - \bar{u}) q^* \mathrm{d}\Gamma \tag{3.2}$$

where $q = \partial u/\partial n$ and $q^* = \partial u^*/\partial n$.

Note that the last term in Equation 3.2 is obtained by weighting the $u = \bar{u}$ conditions on Γ_1 and renders a mixed type statement.

Integrating by parts the left hand side of Equation (3.2) we obtain:

$$-\int_{\Omega} \left\{ \frac{\partial u}{\partial x_k} \frac{\partial u^*}{\partial x_k} \right\} \mathrm{d}\Omega = -\int_{\Gamma_2} \bar{q} u^* \mathrm{d}\Gamma - \int_{\Gamma_1} q u^* \mathrm{d}\Gamma -$$

$$-\int_{\Gamma_1} uq^*\mathrm{d}\Gamma + \int_{\Gamma_1} \bar{u}q^*\mathrm{d}\Gamma \tag{3.3}$$

where the Einstein summation convention for indices has been used. After integrating again the left hand side of Equation (3.3), the following expression can be written

$$\int_{\Omega} u(\nabla^2 u^*)\mathrm{d}\Omega = -\int_{\Gamma_2} \bar{q}u^*\mathrm{d}\Gamma - \int_{\Gamma_1} qu^*\mathrm{d}\Gamma$$

$$+\int_{\Gamma_2} uq^*\mathrm{d}\Gamma + \int_{\Gamma_1} \bar{u}q^*\mathrm{d}\Gamma \tag{3.4}$$

This equation is the starting point for the boundary element method.

Fundamental solution

Our aim is now to find a solution satisfying the Laplace equation. If we assume that a concentrated charge is acting at a point 'i', the governing equation is

$$\nabla^2 u^* + \Delta^i = 0 \tag{3.5}$$

where Δ^i is a Dirac delta function. The solution of this equation is called the fundamental solution. As we have already seen this function has the property that

$$\int_{\Omega} u(\nabla^2 u^* + \Delta^i)\mathrm{d}\Omega = \int_{\Omega} u\nabla^2 u^*\mathrm{d}\Omega + u^i \tag{3.6}$$

where u^i represents the value of the unknowns function u at the point of application of the charge. If Equation (3.5) is satisfied by the fundamental solution,

$$\int_{\Omega} u(\nabla^2 u^*)\mathrm{d}\Omega = -u^i \tag{3.7}$$

Then Equation (3.4) becomes,

$$u^i + \int_{\Gamma_2} uq^*\mathrm{d}\Gamma + \int_{\Gamma_1} \bar{u}q^*\mathrm{d}\Gamma = \int_{\Gamma_2} \bar{q}u^*\mathrm{d}\Gamma + \int_{\Gamma_1} qu^*\mathrm{d}\Gamma \tag{3.8}$$

where

$$q = \frac{\partial u}{\partial n}, \qquad q^* = \frac{\partial u^*}{\partial n}$$

For an isotropic three dimensional medium the fundamental solution of Equation (3.5) is

$$u^* = \frac{1}{4\pi r} \tag{3.9}$$

where r is the distance from the point of application of the delta function to the point under consideration. The three dimensional Laplace equation (3.5) in polar coordinates and taking symmetry into consideration becomes,

$$\frac{\partial^2 u^*}{\partial r^2} + \frac{2}{r}\frac{\partial u^*}{\partial r} = -\Delta^i \tag{3.10}$$

Substituting Equation (3.9) into (3.10) we see that the equation is satisfied for any value of r different from zero. To study the case of $r \equiv 0$ we have to carry out the following integration, in a sphere surrounding the point where the charge is applied,

$$\int_\Omega \nabla^2 u^* \mathrm{d}\Omega = -\int_\Omega \Delta^i \mathrm{d}\Omega = -1 \tag{3.11}$$

We can easily prove that the left hand side of Equation (3.11) is also equal to minus one by writing the following expression

$$\int_\Omega \nabla^2 u^* \mathrm{d}\Omega = \int_\Gamma \frac{\partial u^*}{\partial n} \mathrm{d}\Gamma = \int_\Gamma \frac{\partial u^*}{\partial r} \mathrm{d}\Gamma \tag{3.12}$$

Substituting the fundamental solution (3.9) into (3.12) we obtain,

$$\int_\Gamma \frac{\partial u^*}{\partial r} \mathrm{d}\Gamma = \frac{1}{4\pi} \int_\Gamma \left(-\frac{1}{r^2}\right) \mathrm{d}\Gamma = -\frac{1}{4\pi}\left(\frac{4\pi r^2}{r^2}\right)_\Gamma = -1 \tag{3.13}$$

The result in Equation (3.13) is independent of r and shows that when $r \to 0$ the left hand side of Equation (3.11) is also equal to -1.

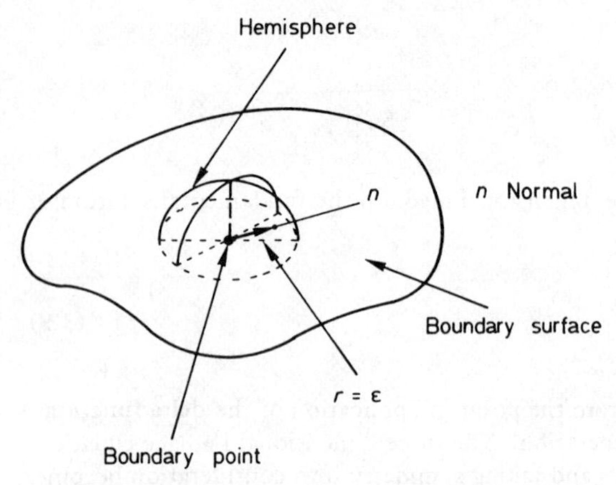

Fig.3.2. Boundary surface assumed to be hemispherical for integration purposes

For two dimensions the fundamental solution for the isotropic case is

$$u^* = \frac{1}{2\pi} \ln \left(\frac{1}{r} \right) \tag{3.14}$$

and the same considerations as above apply.

Equation on the boundary

Equation (3.8) is valid for any point in the domain, but in order to formulate the problem as a boundary technique one needs to take it to the boundary. This will now be done in a simple way. Consider the hemisphere on the boundary of a three dimensional domain as depicted in Figure 3.2. The two-dimensional case can also be analysed in the same way. The boundary point is assumed to be at the centre of the sphere and afterwards the radius 'ϵ' is reduced to zero. The point will then become a boundary point. The boundary is assumed to be smooth and the point to be on Γ_2, but similar considerations apply if the point is on Γ_1.

In order to analyse what is happening consider that the Γ_2 boundary is divided into two parts, i.e.

$$\int_{\Gamma_2} u \frac{\partial u^*}{\partial n} \, d\Gamma = \int_{\Gamma_{(2-\epsilon)}} u \frac{\partial u^*}{\partial n} \, d\Gamma + \int_{\Gamma_\epsilon} u \frac{\partial u^*}{\partial n} \, d\Gamma \tag{3.15}$$

We can now substitute the fundamental solution into the second integral on the right hand side of Equation (3.15) and take it to the limit, i.e. $\epsilon \to 0$. This gives,

$$\lim_{\epsilon\to 0}\left\{\int_{\Gamma_\epsilon} u\frac{\partial u^*}{\partial n}\,d\Gamma\right\} = \lim_{\epsilon\to 0}\left\{-\int_{\Gamma_\epsilon} u\frac{1}{4\pi\epsilon^2}\,d\Gamma\right\} =$$

$$= \lim_{\epsilon\to 0}\{-\tfrac{1}{2}u\} = -\tfrac{1}{2}u \qquad (3.16)$$

Note that as ϵ is now zero the boundary $\Gamma_{(2-\epsilon)}$ again becomes Γ_2. The same result, i.e. $-\tfrac{1}{2}u$, is obtained in two dimensional problems for a smooth boundary. The subdivision can also be introduced for the right hand side terms of Equation (3.8), i.e.

$$\int_\Gamma qu^*d\Gamma$$

but for this case

$$\lim_{\epsilon\to 0}\left\{\int_{\Gamma_\epsilon} q\frac{1}{4\pi\epsilon}\,d\Gamma\right\} = 0$$

and hence this limit does not introduce any new term in Equation (3.8). Substituting Equation (3.16) into Equation (3.8) one has the following equation for a point on the boundary:

$$\tfrac{1}{2}u^i + \int_{\Gamma_2} u\frac{\partial u^*}{\partial n}\,d\Gamma + \int_{\Gamma_1} \bar{u}\frac{\partial u^*}{\partial n}\,d\Gamma = \int_{\Gamma_2} \bar{q}u^*d\Gamma + \int_{\Gamma_1} qu^*d\Gamma \qquad (3.17)$$

The same result would be obtained if instead of a point on Γ_2 one considers a point on the Γ_1 part of the boundary.

In general Equation (3.17) can be written as,

$$\tfrac{1}{2}u^i + \int_\Gamma uq^*d\Gamma = \int_\Gamma qu^*d\Gamma \qquad (3.18)$$

where $\Gamma = \Gamma_1 + \Gamma_2$ and one assumes that $u = \bar{u}$ on Γ_1 and $\partial u/\partial n = q = \bar{q}$ on Γ_2.

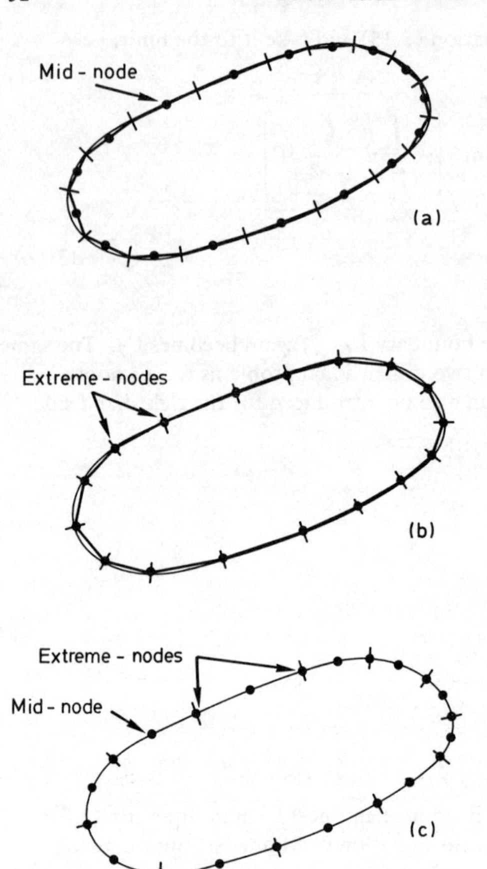

Fig.3.3. Different types of boundary elements; (a) constant elements, (b) linear elements, (c) quadratic elements

3.3 BOUNDARY ELEMENTS

Equation (3.18) will now be applied on the boundary of the domain under consideration. For simplicity only the two-dimensional case will be studied here and its boundary will be divided into n straight line segments or elements (Figure 3.3). The points where the unknown values are considered or 'nodes' are taken to be in the middle of each segment (Figure 3.3(a)) or at the intersection between two elements (Figure 3.3(b)). In addition the boundary does not need to be discretized by straight segments only but one can also use curved elements with little extra work, (Figure 3.3(c)). For the latter case an extra mid-element node is generally included.

The case of having only middle nodes will first be considered (Figure 3.3(a)). The boundary has been discretized into n elements, of which n_1

belong to Γ_1 and n_2 to Γ_2. The values of u and q are assumed to be constant on each element and equal to the value at the mid-node of the element. Equation (3.18) for a given 'i' point becomes in discretized form

$$\tfrac{1}{2}u^i + \sum_{j=1}^{n} u_j \int_{\Gamma_j} q^* \mathrm{d}\Gamma = \sum_{j=1}^{n} q_j \int_{\Gamma_j} u^* \mathrm{d}\Gamma \tag{3.19}$$

This equation applies for a particular node 'i'. The terms

$$\int_{\Gamma_j} q^* \mathrm{d}\Gamma$$

relate the 'i' node with the segment 'j' over which the integral is carried out. We shall call these integrals $\hat{H}_{ij}$. The integrals on the right hand side will be called G_{ij}. Hence one has

$$\tfrac{1}{2}u^i + \sum_{j=1}^{n} u_j \hat{H}_{ij} = \sum_{j=1}^{n} q_j G_{ij} \tag{3.20}$$

The above integrals are easy to calculate analytically for the constant element case but for higher order elements they become more difficult to evaluate. For generality the integrals were calculated numerically for all segments except the one corresponding to the node under consideration.

Equation (3.20) relates the value of u at mid-node 'i' with the value of u and q at all the nodes on the boundary, including 'i'.

One can write Equation (3.20) for each 'i' node, obtaining n equations. Let us now call

$$H_{ij} = \hat{H}_{ij}, \quad \text{when } i \neq j$$

$$H_{ij} = \hat{H}_{ij} + \tfrac{1}{2}, \quad \text{when } i = j \tag{3.21}$$

Equation (3.20) can then be written as:

$$\sum_{j=1}^{n} H_{ij} u_j = \sum_{j=1}^{n} G_{ij} q_j \tag{3.22}$$

The whole set of equations for the n nodes can now be expressed in matrix form as:

$$\mathbf{H}\,\mathbf{U} = \mathbf{G}\,\mathbf{Q} \tag{3.23}$$

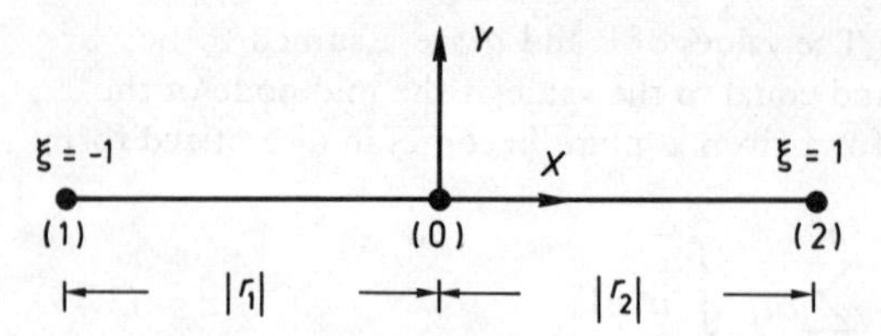

Fig.3.4. Linear element coordinates

Note that n_1 values of u and n_2 values of q are known, hence one has a set of n unknowns in formula (3.23). Reordering the equations in such a way that all the unknowns are on the left hand side, one can write:

$$\mathbf{A}\,\mathbf{X} = \mathbf{F} \tag{3.24}$$

where $\mathbf{X}$ is the vector of unknowns u and q.

Once the values of u and q on the whole boundary are known one can calculate the value of u at any interior point using Equation (3.8), which after discretization becomes:

$$u^i = \sum_{j=1}^{n} q_j G_{ij} - \sum_{j=1}^{n} u_j \hat{H}_{ij} \tag{3.25}$$

The internal fluxes $q_x (q_x = \partial u/\partial x)$ or $q_y (q_y = \partial u/\partial y)$ can be calculated by differentiating Equation (3.8), i.e.

$$(q_x)^i = \int_\Gamma q \frac{\partial u^*}{\partial x}\, \mathrm{d}\Gamma - \int_\Gamma u \frac{\partial q^*}{\partial x}\, \mathrm{d}\Gamma$$

$$(q_y)^i = \int_\Gamma q \frac{\partial u^*}{\partial y}\, \mathrm{d}\Gamma - \int_\Gamma u \frac{\partial q^*}{\partial y}\, \mathrm{d}\Gamma \tag{3.26}$$

Integration

The integrals can be calculated using a 4-point Gauss quadrature rule for all segments* except the one corresponding to the node under consideration. For this particular case the $\hat{H}_{ii}$ is zero (due to the orthogonality of r and n) and the G_{ii} integral can be calculated analytically.

$$G_{ii} = \int_{\Gamma_i} u^* \mathrm{d}\Gamma = \frac{1}{2\pi} \int_{\Gamma_i} \ln\left(\frac{1}{r}\right) \mathrm{d}\Gamma \tag{3.27}$$

* See Appendix for numerical integration formulae.

We can now use the ξ coordinate (Figure 3.4) and have,

$$\frac{1}{2\pi}\int_{(1)}^{(2)} \ln\left(\frac{1}{r}\right) d\Gamma = \frac{1}{\pi}\int_{(0)}^{(2)} \ln\left(\frac{1}{r}\right) dr \tag{3.28}$$

( ) indicates that the number between brackets corresponds to the point number not the distance.

Transforming coordinates (note $r = \xi|r_1|$, where $|r_1| = |r_2|$) one obtains:

$$\frac{1}{\pi}\int_{(0)}^{(2)} \ln\left(\frac{1}{r}\right) dr = \frac{1}{\pi}|r_1|\left\{\ln\frac{1}{|r_1|} + \int_0^1 \ln\left(\frac{1}{\xi}\right) d\xi\right\} \tag{3.29}$$

where the last integral can easily be calculated, i.e. equals 1.
Hence Equation (3.27) becomes,

$$G_{ii} = \frac{1}{\pi}|r_1|\left\{\ln\frac{1}{|r_1|} + 1\right\} \tag{3.30}$$

For more complex cases a special logarithmically weighted numerical integration formula can be used. The formula is due to Stroud and Secrest and is given in Appendix 1. The integration implies that,

$$\int_0^1 \ln\left(\frac{1}{r}\right) f(r)\,dr \simeq \sum_{i=1}^{k} w_i f(r_i) \tag{3.31}$$

Linear elements

The variation of u and q will now be assumed to be linear within each element. The nodes are considered to be at the intersection between two straight elements such as those shown in Fig.3.3(b) and marked as extreme-nodes.

Equation (3.18) can be written for the n elements;

$$c^i u^i + \sum_{j=1}^{n}\int_{\Gamma_j} uq^*\,d\Gamma = \sum_{j=1}^{n}\int_{\Gamma_j} qu^*\,d\Gamma \tag{3.32}$$

Note that contrary to Equation (3.19) we cannot now take u_j and q_j out of

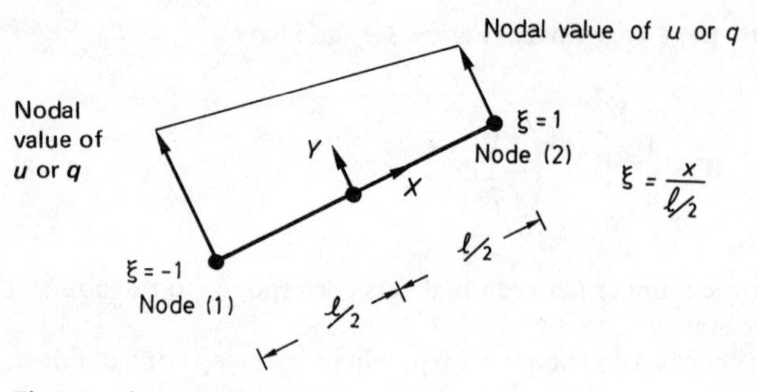

Fig.3.5. Linear element

the integral as they vary linearly within the element. $c^i = ½$ for smooth boundaries but a way of finding its value for non-smooth boundaries will be shown.

Consider an arbitrary segment such as the one shown in Figure 3.5. The values of u and q at any point of the element can be defined in terms of their nodal values and the linear interpolation functions ϕ_1 and ϕ_2 such that, for the case of both variables having a linear variation we can write

$$u(\xi) = \phi_1 u_1 + \phi_2 u_2 = [\phi_1 \; \phi_2] \begin{Bmatrix} u_1 \\ u_2 \end{Bmatrix}$$

$$q(\xi) = \phi_1 q_1 + \phi_2 q_2 = [\phi_1 \; \phi_2] \begin{Bmatrix} q_1 \\ q_2 \end{Bmatrix} \tag{3.33}$$

where ξ is the dimensionless coordinate $\xi = 2x/l$ and ϕ_1, ϕ_2 are given by

$$\phi_1 = ½(1 - \xi), \quad \phi_2 = ½(1 + \xi) \tag{3.34}$$

The integrals along a 'j' element in Equation (3.32) become for the left hand side,

$$\int_{\Gamma_j} uq^* \mathrm{d}\Gamma = \int_{\Gamma_j} [\phi_1 \; \phi_2]\, q^* \mathrm{d}\Gamma \begin{Bmatrix} u_1 \\ u_2 \end{Bmatrix} = [h_{i1} \; h_{i2}] \begin{Bmatrix} u_1 \\ u_2 \end{Bmatrix} \tag{3.35}$$

where for each 'j' element we have two components,

$$h_{i1} = \int_{\Gamma_j} \phi_1 q^* \mathrm{d}\Gamma, \quad h_{i2} = \int_{\Gamma_j} \phi_2 q^* \mathrm{d}\Gamma$$

For the right hand side we have,

$$\int_{\Gamma_j} q u^* d\Gamma = \int_{\Gamma_j} [\phi_1 \; \phi_2] \, u^* d\Gamma \begin{Bmatrix} q_1 \\ q_2 \end{Bmatrix} = [g_{i1} \; g_{i2}] \begin{Bmatrix} q_1 \\ q_2 \end{Bmatrix} \tag{3.36}$$

where

$$g_{i1} = \int_{\Gamma_j} \phi_1 u^* d\Gamma, \quad g_{i2} = \int_{\Gamma_j} \phi_2 u^* d\Gamma.$$

Substituting Equations (3.35) and (3.36) for all 'j' elements into (3.32) one obtains the following equation for node 'i'

$$c^i u^i + [\hat{H}_{i1} \; \hat{H}_{i2} \ldots \hat{H}_{in}] \begin{Bmatrix} u_1 \\ u_2 \\ \cdot \\ \cdot \\ \cdot \\ u_n \end{Bmatrix} = [G_{i1} \; G_{i2} \ldots G_{in}] \begin{Bmatrix} q_1 \\ q_2 \\ \cdot \\ \cdot \\ \cdot \\ q_n \end{Bmatrix} \tag{3.37}$$

where $\hat{H}_{ij}$ is equal to the h_{i1} term of element 'j' plus h_{i2} term of element 'j-1' and similarly for G_{ij}. Hence formula (3.37) represents the assembled equation for node 'i'. Note the simplicity of this approach by comparison with finite elements. Equation (3.37) can be written as,

$$c^i u^i + \sum_{j=1}^{n} \hat{H}_{ij} u_j = \sum_{j=1}^{n} G_{ij} q_j \tag{3.38}$$

where 'j' defines the nodes in between elements (extreme-nodes in Figure 3.3). Similarly, as it was previously shown (Equation 3.22), this equation can be written as

$$\sum_{j=1}^{n} H_{ij} u_j = \sum_{j=1}^{n} G_{ij} q_j \tag{3.39}$$

and the whole set in matrix form becomes

$$\mathbf{H} \, \mathbf{U} = \mathbf{G} \, \mathbf{Q} \tag{3.40}$$

Unless the surface is smooth at the point 'i' the value of $-\frac{1}{2}$ obtained in Equation (3.16) is not valid. However, one can always calculate the diagonal terms of **H** by the fact that when a uniform potential is applied over the whole boundary, the normal derivatives (q values) must be zero. Hence Equation (3.40) becomes,

$$\mathbf{H}\,\mathbf{U} = \mathbf{0} \tag{3.41}$$

Thus the sum of all the elements of **H** in any row ought to be zero, and the value of the coefficient on the diagonal can be easily calculated once the off-diagonal coefficients are all known, i.e.

$$h_{ii} = -\sum_{\substack{j=1\\ j\neq i}}^{n} h_{ij}$$

In this way one need not calculate explicitly the value of c^i.

3.4 SIMPLE COMPUTER PROGRAM

We will now describe a simple FORTRAN computer program for the solution of isotropic potential problems using constant elements, i.e. elements with constant u and q and a mid-element node only (Figure 3.3(a)).

The program requires less steps than a finite element program as it is now unnecessary to have a special assembler subroutine. The number of unknowns is substantially smaller as only nodes on the boundary are required. Figure 3.6 compares the main steps of the two methods. Not only is the number of steps reduced for boundary elements but the data input is very much easier than for finite elements. Internal results are computed only at required points and not everywhere in the domain, as is the case with finite elements.

Main program and data structure

The macro flow diagram for the boundary element program can be seen in Figure 3.7. The main program defines the maximum dimensions of the system of equations (or boundary nodes) which in this case is 40. It also allocates the input channel 5 and the output channel 6 for the Format statements. It calls the five following subroutines,

- INPUT: Reads the program input.
- FMAT: Forms the two matrices **H** and **G** and rearranges them according to the boundary conditions to form the matrix A of Equation (3.24).
- SLNPD: Solution of the system of equations by Gauss elimination.

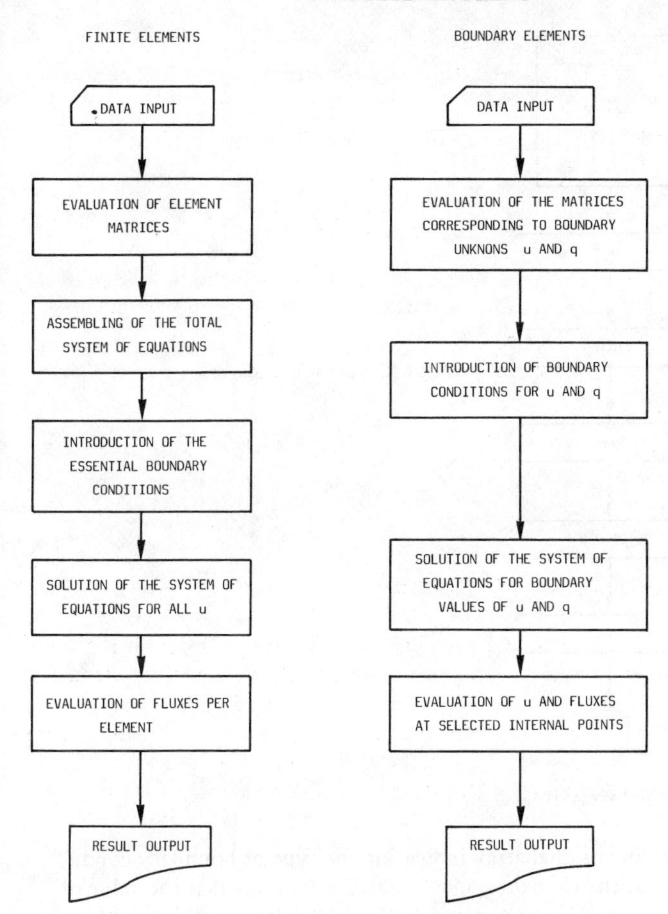

Fig.3.6. Finite element program versus boundary element program

INTER: Reorders the unknown vector and computes the values of the potential at the selected internal points (values of fluxes are not computed here but could easily be included).

OUTPT: Outputs the results.

The general integer variables used by the program, together with their meaning are given below,

N: Number of boundary elements (equal to number of nodes in this case).

L: Number of internal points where the function is calculated.

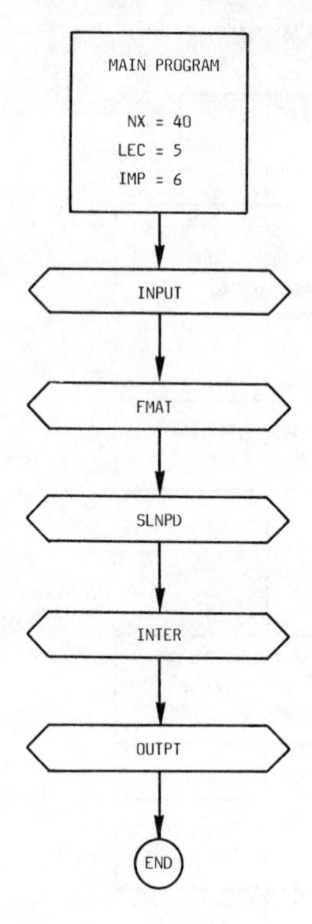

Fig.3.7. Macro-flow diagram

The only integer array is,

KODE: One dimensional array indicating the type of boundary conditions at the element nodes. KODE = 0 means that the value of the potential is known and KODE = 1 means that the value of q is known at the boundary node.

There are several real arrays to store both data and results.

X: One dimensional array of x coordinates extreme point of boundary elements.

Y: One dimensional array of y coordinates extreme point of boundary elements.

XM: x coordinates of the nodes. XM(J) contains the x coordinate of node J.

YM: y coordinate of the nodes. YM(J) contains the y coordinate of node J.

G: Matrix defined in Equation (3.23). After application of boundary conditions the matrix A is stored in the same location.

H: Matrix defined in Equation (3.23).

FI: Prescribed value of boundary conditions. FI(J) contains the prescribed value of the condition at node J. If KODE = 0 it means that the potential is prescribed and if KODE = 1 that the q is given.

DFI: Right hand side vector in Equation (3.24). After solution it contains the values of the u's and q's unknowns.

CX: x coordinate for internal point where the value of u is required.

CY: y coordinate for internal point where the value of u is required.

SOL: Vector of the potential values for internal points.

```
C
C  PROGRAM 1
C
C
C  PROGRAM FOR SOLUTION OF TWO DIMENSIONAL POTENTIAL PROBLEMS
C  BY THE B.I.E. METHOD WITH CONSTANT ELEMENTS
C
      COMMON N,L,LEC,IMP
      DIMENSION X(41), Y(41), XM(40), YM(40), G(40, 40), FI(40), DFI(40)
      DIMENSION KODE(40), CX(40), CY(40), SOL(40), H(40, 40)
C
C  INITIALIZATION OF PROGRAM PARAMETERS
C  NX = MAXIMUM DIMENSION OF THE SYSTEM OF EQUATIONS.
C
      NX = 40
C
C  ASSIGN DATA SET NUMBERS FOR INPUT, LEC, AND OUTPUT, IMP
C
      LEC = 5
      IMP = 6
C
C  INPUT
C
      CALL INPUT (CX, CY, X, Y, KODE, FI)
C
C  FORM SYSTEM OF EQUATIONS
C
      CALL FMAT(X, Y, XM, YM, G, H, FI, DFI, KODE, NX)
C
C  SOLUTION OF THE SYSTEM OF EQUATIONS
C
      CALL SLNPD (G, DFI, D, N, NX)
C
C  COMPUTE THE POTENTIAL VALUES IN INTERNAL POINTS
C
      CALL INTER (FI, DFI, KODE, CX, CY, X, Y, SOL)
C
C  OUTPUT
C
      CALL OUTPT (XM, YM, FI, DFI, CX, CY, SOL)
      STOP
      END
```

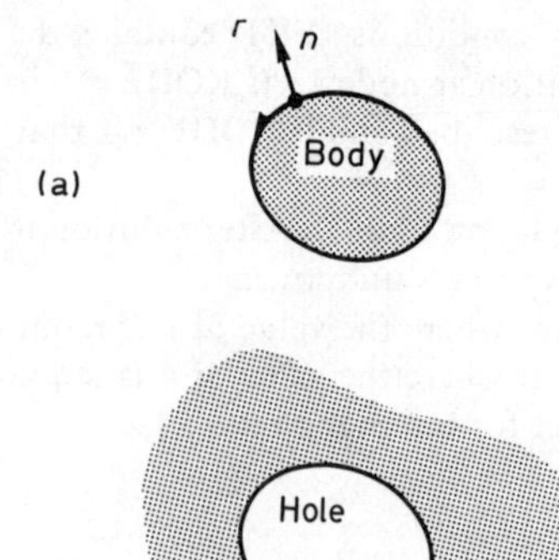

Fig.3.8. (a) Numbering scheme for (a) external surface (b) internal surface

Subroutine input

All the input required by the program is read in the program INPUT. The input data will consist of the following group of cards.

(1) *Title Card* One card containing the name of the program, FORMAT 18A4.

(2) *Basic Parameters Card* One card containing the number of boundary elements and the number of internal points where the function is required. FORMAT 2I5.

(3) *Internal Points Coordinates Cards* As many cards as internal nodes are required each with the xy nodal coordinates FORMAT 2F10.4.

(4) *Extreme Points of Boundary Elements Cards* Each card defines the coordinates of the extreme of an element, read in counter-clockwise direction for the case shown in Figure 3.8(a) and in clockwise direction for 3.8(b). FORMAT 2F10.4.

(5) *Boundary Condition Cards* As many cards as nodes giving the values of the potential at the node if KODE = 0 or the value of the potential derivative if KODE = 1.

This subroutine prints the title, the basic parameters, the extreme points of the boundary elements and the boundary conditions. The internal point coordinates are printed in the OUTPT subroutine.

```
      SUBROUTINE INPUT (CX, CY, X, Y, KODE, FI)
C
C  PROGRAM 2
C
      COMMON N, L, LEC, IMP
      DIMENSION CX(1), CY(1), X(1), Y(1), KODE(1), FI(1), TITLE(18)
```

```
C
C N  =  NUMBER OF BOUNDARY ELEMENTS
C L  =  NUMBER OF INTERNAL POINTS WHERE THE FUNCTION IS
C       CALCULATED
C
        WRITE (IMP, 100)
  100   FORMAT (' ', 120(' * '))
C
C  READ NAME OF THE JOB
C
        READ (LEC, 150) TITLE
  150   FORMAT (18A4)
        WRITE (IMP, 250) TITLE
  250   FORMAT (25X, 18A4)
C
C  READ BASIC PARAMETERS
C
        READ (LEC, 200) N, L
  200   FORMAT (2I5)
        WRITE (IMP, 300) N, L
  300   FORMAT (//'DATA'//2X, 'NUMBER OF BOUNDARY ELEMENTS=',
      1 I3/2X, 'NUMBER OF INTERNAL POINTS WHERE THE FUNCTION IS
      2 CALCULATED=', I3)
C
C  READ INTERNAL POINTS COORDINATES
C
        DO 1 I = 1, L
    1   READ (LEC, 400) CX(I), CY(I)
  400   FORMAT (2F10.4)
C
C  READ COORDINATES OF EXTREME POINTS OF THE BOUNDARY
C  ELEMENTS IN ARRAY X AND Y
C
        WRITE (IMP, 500)
  500   FORMAT (//2X, 'COORDINATES OF THE EXTREME POINTS OF THE
      1 BOUNDARY ELEMENTS', //4X, 'POINT', 10X, 'X', 18X, 'Y')
        DO 10 I = 1, N
        READ (LEC, 600) X(I), Y(I)
  600   FORMAT (2F10.4)
   10   WRITE (IMP, 700) I, X(I), Y(I)
  700   FORMAT (5X, I3, 2(5X, E14.7))
C
C  READ BOUNDARY CONDITIONS
C  FI(I) = VALUE OF THE POTENTIAL IN THE NODE I IF KODE = 0,
C  VALUE OF THE POTENTIAL DERIVATIVE IF KODE = 1.
C
        WRITE (IMP, 800)
  800   FORMAT (//2X, 'BOUNDARY CONDITIONS'//5X, 'NODE', 6X, 'CODE',
      1 5X, 'PRESCRIBED VALUE')
        DO 20 I = 1, N
        READ (LEC, 900) KODE (I), FI(I)
  900   FORMAT (I5, F10.4)
   20   WRITE (IMP, 950) I, KODE(I), FI(I)
  950   FORMAT (5X, I3, 8X, I1, 8X, E14.7)
        RETURN
        END
```

Forming the total matrix and its right hand side vector

The subroutine FMAT computes the **G** and **H** matrices of Equation (3.23). Their elements are computed by using the subroutines INTE and INLO.

INTE: This subroutine computes the **H** and **G** matrix elements by means of numerical integration along the boundary elements. It calculates all elements except those on the diagonal.

INLO: Calculates the diagonal elements of **G** matrix, given by Equation (3.30).

The diagonal element of the matrix **H** is $\hat{H}_{ii} + \frac{1}{2}$ (Equation 3.21) and is equal to ½. Note that as the fundamental solution has been taken to be $\ln(1/r)$ in the program, all the G and H terms appear multiplied by 2π.

Next the system of equations is rearranged to form the **A** matrix, (now in **G**) and the right hand side vector **F** (now DFI) of Equation (3.24).

```
      SUBROUTINE FMAT (X, Y, XM, YM, G, H, FI, DFI, KODE, NX)
C
C  PROGRAM 3
C
C  THIS SUBROUTINE COMPUTES G AND H MATRICES AND FORM THE
C  SISTEM AX = F
C
      COMMON N, L, LEC, IMP
      DIMENSION X(1), Y(1), XM(1), YM(1), G(NX, NX), H(NX, NX), FI(1),KODE(1)
      DIMENSION DFI(1)
C
C  COMPUTE THE MID-POINT COORDINATES AND STORE IN ARRAY XM AND YM
C
      X(N + 1) = X(1)
      Y(N + 1) = Y(1)
      DO 10  I = 1, N
      XM(I) = (X(I) + X(I + 1))/2
   10 YM(I) = (Y(I) + Y(I + 1))/2
C
C  COMPUTE G AND H MATRICES
C
      DO 30  I = 1, N
      DO 30  J = 1, N
      IF(I - J)20, 25, 20
   20 CALL INTE(XM(I), YM(I), X(J), Y(J), X(J + 1), Y(J + 1), H(I, J), G(I, J))
      GO TO 30
   25 CALL INLO(X(J), Y(J), X(J + 1), Y(J + 1), G(I, J))
      H(I, J) = 3.1415926
   30 CONTINUE
C
C  ARRANGE THE SYSTEM OF EQUATIONS READY TO BE SOLVED
C
      DO 50 J = 1, N
      IF(KODE(J)) 50, 50, 40
```

```
   40 DO 50  I = 1, N
      CH = G(I, J)
      G(I, J) = -H(I, J)
      H(I, J) = -CH
   50 CONTINUE
C
C  DFI ORIGINALLY CONTAINS THE INDEPENDENT COEFFICIENTS
C  AFTER SOLUTION IT WILL CONTAIN THE VALUES OF THE SYSTEM UNKNOWNS
C
      DO 60  I = 1, N
      DFI(I) = 0.
      DO 60  J = 1, N
      DFI(I) = DFI(I) + H(I, J) * FI(J)
   60 CONTINUE
      RETURN
      END
```

Subroutine INTE

The values of the off-diagonal coefficients of **H** and **G** are calculated using a 4-point Gauss integration formula (see Appendix 1).

The variable DIST is the distance from the point under consideration to the boundary element as shown in Figure 3.9. If its direction is the same as that of the normal it is defined as positive, otherwise it is negative.

The **G** and **H** terms will be of the type

$$\mathbf{G} = \sum_{i=1}^{4} \ln \frac{1}{(\mathrm{RA})_i} \, w_i \, \frac{\sqrt{(\mathrm{X1} - \mathrm{X2})^2 + (\mathrm{Y1} - \mathrm{Y2})^2}}{2}$$

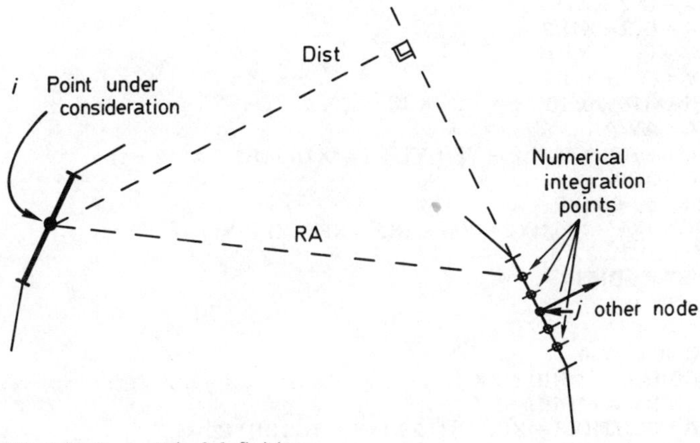

Fig.3.9. Geometrical definitions

$$H = \sum_{i=1}^{4} \frac{d}{dn}\left\{\ln \frac{1}{(RA)_i}\right\} w_i \frac{\sqrt{(X1 - X2)^2 + (Y1 - Y2)^2}}{2}$$

$$= \sum_{i=1}^{4} -\frac{1}{(RA)_i^2} (DIST)\, w_i \frac{\sqrt{(X1 - X2)^2 + (Y1 - Y2)^2}}{2}$$

where X and Y are the coordinates of the extreme points.

The last part of the terms is a consequence of the change of coordinates from X, Y to the dimensionless coordinate ξ.

```
      SUBROUTINE INTE(XP, YP, X1, Y1, X2, Y2, H, G)
C
C  PROGRAM 4
C
C  THIS SUBROUTINE COMPUTES THE VALUES OF THE H AND G MATRIX
C  OFF DIAGONAL ELEMENTS BY MEANS OF NUMERICAL INTEGRATION
C  ALONG THE BOUNDARY ELEMENTS
C
C  DIST = DISTANCE FROM THE POINT UNDER CONSIDERATION TO THE
C  BOUNDARY ELEMENTS
C  RA = DISTANCE FROM THE POINT UNDER CONSIDERATION TO THE
C  INTEGRATION POINTS IN THE BOUNDARY ELEMENTS
C
      DIMENSION XCO(4), YCO(4), GI(4), OME(4)
      GI(1) = 0.86113631
      GI(2) = -GI(1)
      GI(3) = 0.33998104
      GI(4) = -GI(3)
      OME(1) = 0.34785485
      OME(2) = OME(1)
      OME(3) = 0.65214515
      OME(4) = OME(3)
      AX = (X2 - X1)/2
      BX = (X2 + X1)/2
      AY = (Y2 - Y1)/2
      BY = (Y2 + Y1)/2
      IF(AX)10, 20, 10
   10 TA = AY/AX
      DIST = ABS((TA*XP - YP + Y1 - TA*X1)/SQRT(TA**2 + 1)
      GO TO 30
   20 DIST = ABS (XP - X1)
   30 SIG = (X1 - XP)*(Y2 - YP) - (X2 - XP)*(Y1 - YP)
      IF(SIG) 31, 32, 32
   31 DIST = -DIST
   32 G = 0.
      H = 0.
      DO 40 I = 1, 4
      XCO(I) = AX*GI(I) + BX
      YCO(I) = AY*GI(I) + BY
      RA = SQRT((XP - XCO(I))**2 + (YP - YCO(I))**2)
      G = G + ALOG(1/RA)*OME(I)*SQRT(AX**2 + AY**2)
```

```
40 H = H – (DIST*OME(I)*SQRT(AX**2 + AY**2)/RA**2)
   RETURN
   END
```

Subroutine INLO

Calculates the diagonal elements of **H** using formula (3.30) multiplied by 2π, i.e.

$$G_{ii} = 2\ |r| \left\{ \ln \frac{1}{|r|} + 1 \right\}$$

```
      SUBROUTINE INLO(X1, Y1, X2, Y2, G)
C
C PROGRAM 5
C
C THIS SUBROUTINE COMPUTES THE VALUES OF THE DIAGONAL
C ELEMENTS OF THE G MATRIX
C
      AX = (X2 – X1)/2
      AY = (Y2 – Y1)/2
      SR = SQRT(AX**2 + AY**2)
      G = 2*SR*(ALOG(1/SR) + 1)
      RETURN
      END
```

Subroutine SLNPD

This is a standard subroutine given in reference 11 that can solve non positive definite matrix using Gauss elimination. If the matrix **A** has a zero in the diagonal it will interchange row, deciding that the system matrix is singular only when no row interchange will produce a non zero diagonal coefficient, for a given row position.

The result is a vector with the unknown boundary potentials and derivatives and is overwritten in DFI.

```
      SUBROUTINE SLNPD(A, B, D, N, NX)
C
C PROGRAM 6
C
C SOLUTION OF LINEAR SYSTEMS OF EQUATIONS
C BY THE GAUSS ELIMINATION METHOD PROVIDING
C FOR INTERCHANGING ROWS WHEN ENCOUNTERING A
C ZERO DIAGONAL COEFFICIENT
C
C A : SYSTEM MATRIX
C B : ORIGINALLY IT CONTAINS THE INDEPENDENT
C     COEFFICIENTS. AFTER SOLUTION IT CONTAINS THE
C     VALUES OF THE SYSTEM UNKNOWNS.
```

```
C
C  N : ACTUAL NUMBER OF UNKNOWNS
C  NX : ROW AND COLUMN DIMENSION OF A
C
      DIMENSION A(NX, NX), B(NX)
      N1 = N - 1
      DO 100   K = 1, N1
      K1 = K + 1
      C = A(K, K)
      IF(ABS(C) - 0.000001)1, 1, 3
    1 DO 7 J = K1, N
C
C  TRY TO INTERCHANGE ROWS TO GET NON ZERO DIAGONAL COEFFICIENT
C
      IF(ABS(A(J, K)) - 0.000001)7, 7, 5
    5 DO 6 L = K, N
      C = A(K, L)
      A(K, L) = A(J, L)
    6 A(J, L) = C
      C = B(K)
      B(K) = B(J)
      B(J) = C
      C = A(K, K)
      GO TO 3
    7 CONTINUE
    8 WRITE (6, 2) K
    2 FORMAT('**** SINGULARITY IN ROW', I5)
      D = 0.
      GO TO 300
C
C  DIVIDE ROW BY DIAGONAL COEFFICIENT
C
    3 C = A(K, K)
      DO 4 J = K1, N
    4 A(K, J) = A(K, J)/C
      B(K) = B(K)/C
C
C  ELIMINATE UNKNOWN X(K) FROM ROW I
C
      DO 10  I = K1, N
      C = A(I, K)
      DO 9  J = K1, N
    9 A(I, J) = A(I, J) - C*A(K, J)
   10 B(I) = B(I) - C*B(K)
  100 CONTINUE
C
C  COMPUTE LAST UNKNOWN
C
      IF (ABS(A(N, N)) - 0.000001) 8, 8, 101
  101 B(N) = B(N)/A(N, N)
C
C  APPLY BACKSUBSTITUTION PROCESS TO COMPUTE REMAINING UNKNOWNS
C
      DO 200 L = 1, N1
      K = N - L
      K1 = K + 1
```

```
      DO 200  J = K1, N
  200 B(K) = B(K) – A(K, J)*B(J)
C
C COMPUTE VALUE OF DETERMINANT
C
      D = 1.
      DO 250 I = 1, N
  250 D = D*A(I, I)
  300 RETURN
      END
```

Computing the values of potentials at internal points

Subroutine INTER reorders FI (boundary condition vector) and DFI (unknown vector) in such a way that all the values of the potential are stored in FI and all the values of the derivatives in DFI.

This subroutine also computes the potential values for the internal points using formula (3.25). Note that because all the H and G terms appear multiplied by 2π the solution for the internal points is eventually multiplied by 2π.

```
      SUBROUTINE INTER(FI, DFI, KODE, CX, CY, X, Y, SOL)
C
C  PROGRAM 7
C
C  THIS SUBROUTINE COMPUTES THE POTENTIAL VALUE FOR INTERNAL
C  POINTS.
      COMMON N, L, LEC, IMP
      DIMENSION FI(1), DFI(1), KODE(1), CX(1), CY(1), X(1), Y(1), SOL(1)
C
C  REORDER FI AND DFI ARRAY TO PUT ALL THE VALUES OF THE POTENTIAL
C  IN FI AND ALL THE VALUES OF THE DERIVATIVE IN DFI
C
      DO 20 I = 1, N
      IF (KODE (I)) 20, 20, 10
   10 CH = FI(I)
      FI(I) = DFI(I)
      DFI(I) = CH
   20 CONTINUE
C
C  COMPUTE THE POTENTIAL VALUES FOR INTERNAL POINTS
C
      DO 40  K = 1, L
      SOL (K) = 0.
      DO 30  J = 1, N
      CALL INTE(CX(K), CY(K), X(J), Y(J), X(J + 1), Y(J + 1), A, B)
   30 SOL(K) = SOL(K) + DFI(J)*B – FI(J)*A
   40 SOL(K) = SOL(K)/(2*3.1415926)
      RETURN
      END
```

Output routine

The results are output in the subroutine OUTPT. This routine lists the coordinates of the boundary nodes and the values of potential and potential derivative and the values of potential at the required internal points.

```
      SUBROUTINE OUTPT(XM, YM, FI, DFI, CX, CY, SOL)
C
C  PROGRAM 8
C
      COMMON N, L, LEC, IMP
      DIMENSION XM(1), YM(1), FI(1), DFI(1), CX(1), CY(1), SOL(1)
      WRITE (IMP, 100)
  100 FORMAT (' ', 120('*')//1X, 'RESULTS'//2X, 'BOUNDARY NODES'//16X,
     1 'X', 23X, 'Y', 19X, 'POTENTIAL', 10X, 'POTENTIAL DERIVATIVE'/)
      DO 10 I = 1, N
   10 WRITE (IMP, 200) XM(I), YM(I), FI(I), DFI(I)
  200 FORMAT (4(10X, E14.7))
      WRITE (IMP, 300)
  300 FORMAT(//, 2X, 'INTERNAL POINTS', //11X, 'X', 18X, 'Y', 14X,
     1 'POTENTIAL', /)
      DO 20 K = 1, L
   20 WRITE (IMP, 400) CX(K), CY(K), SOL(K)
  400 FORMAT (3(5X, E14.7))
      WRITE (IMP, 500)
  500 FORMAT (' ', 120('*'))
      RETURN
      END
```

An example of utilization of the program

As an illustration of the application of the program let us consider a heat flow example for the domain shown in Figure 3.10. The number of boundary elements is 12 and there are 5 internal points where this function is calculated.

The input cards are as follows:

```
HEAT FLOW EXAMPLE
12      5
        2.          2.  }
        4.          2.  }
        4.          4.  }  Internal Points Coordinates
        2.          4.  }
        3.          3.  }
        0.          0.  }
        2.          0.  }
        4.          0.  }
        6.          0.  }
        6.          2.  }
        6.          4.  }
        6.          6.  }  Coordinates of Extreme Points of Boundary
        4.          6.  }  Elements
        2.          6.  }
        0.          6.  }
        0.          4.  }
        0.          2.  }
```

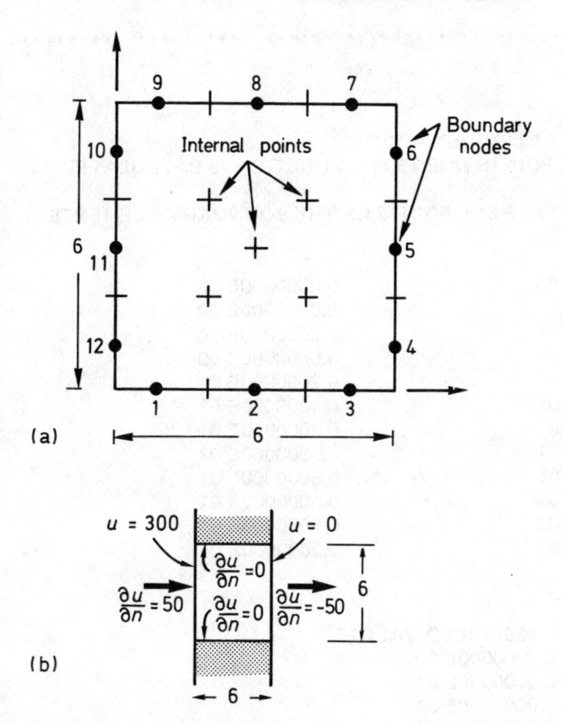

Fig.3.10. Simple potential problem; (a) geometrical definitions, (b) boundary conditions

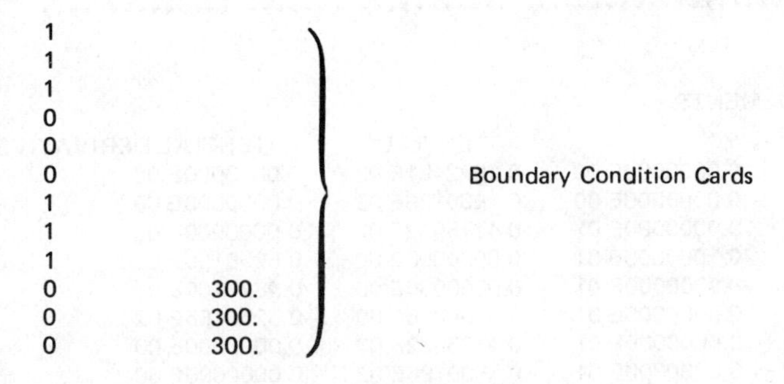

```
1
1
1
0
0
0                      Boundary Condition Cards
1
1
1
0        300.
0        300.
0        300.
```

This gives the following printout:

****

HEAT FLOW EXAMPLE

DATA

NUMBER OF BOUNDARY ELEMENTS = 12
NUMBER OF INTERNAL POINTS WHERE THE FUNCTION IS CALCULATED = 5

COORDINATES OF THE EXTREME POINTS OF THE BOUNDARY ELEMENTS

POINT	X	Y
1	0.0000000E 00	0.0000000E 00
2	0.2000000E 01	0.0000000E 00
3	0.4000000E 01	0.0000000E 00
4	0.6000000E 01	0.0000000E 00
5	0.6000000E 01	0.2000000E 01
6	0.6000000E 01	0.4000000E 01
7	0.6000000E 01	0.6000000E 01
8	0.4000000E 01	0.6000000E 01
9	0.2000000E 01	0.6000000E 01
10	0.0000000E 00	0.6000000E 01
11	0.0000000E 00	0.4000000E 01
12	0.0000000E 00	0.2000000E 01

BOUNDARY CONDITIONS

NODE	CODE	PRESCRIBED VALUE
1	1	0.0000000E 00
2	1	0.0000000E 00
3	1	0.0000000E 00
4	0	0.0000000E 00
5	0	0.0000000E 00
6	0	0.0000000E 00
7	1	0.0000000E 00
8	1	0.0000000E 00
9	1	0.0000000E 00
10	0	0.3000000E 03
11	0	0.3000000E 03
12	0	0.3000000E 03

****

RESULTS

BOUNDARY ELEMENTS

X	Y	POTENTIAL	POTENTIAL DERIVATIVE
0.1000000E 01	0.0000000E 00	0.2522491E 03	0.0000000E 00
0.3000000E 01	0.0000000E 00	0.1500186E 03	0.0000000E 00
0.5000000E 01	0.0000000E 01	0.4775032E 02	0.0000000E 00
0.6000000E 01	0.1000000E 01	0.0000000E 00	−0.5296155E 02
0.6000000E 01	0.3000000E 01	0.0000000E 00	−0.4877100E 02
0.6000000E 01	0.5000000E 01	0.0000000E 00	−0.5296155E 02
0.5000000E 01	0.6000000E 01	0.4775032E 02	0.0000000E 00
0.3000000E 01	0.6000000E 01	0.1500186E 03	0.0000000E 00
0.1000000E 01	0.6000000E 01	0.2522491E 03	0.0000000E 00
0.0000000E 00	0.5000000E 01	0.3000000E 03	0.5295944E 02
0.0000000E 00	0.3000000E 01	0.3000000E 03	0.4873693E 02
0.0000000E 00	0.1000000E 01	0.3000000E 03	0.5296944E 02

INTERNAL POINTS

X	Y	POTENTIAL
0.2000000E 01	0.2000000E 01	0.2002798E 03
0.4000000E 01	0.2000000E 01	0.9973988E 02
0.4000000E 01	0.4000000E 01	0.9973988E 02
0.2000000E 01	0.4000000E 01	0.2002798E 03
0.3000000E 01	0.3000000E 01	0.1500093E 03

3.5 COMPUTER PROGRAM FOR LINEAR ELEMENTS

The theory for linear elements was developed in Section 3.3, but although the programming steps are basically the same most of the subroutines require some modifications.

The following routines are the same in both programs, INPUT, SLNPD.

Main program

The main program is similar to the one in Section 2.4 but does not need the XM, YM arrays with the coordinates of the mid-element nodes, as now the nodes are at the inter-element junction. XM, YM does not need to be included in the argument of subroutine FMAT.

The listing is as follows:

```
C
C  PROGRAM 9
C
C  PROGRAM FOR SOLUTION OF TWO DIMENSIONAL POTENTIAL PROBLEMS
C  BY THE B.I.E. METHOD.  LINEAR VARIATION ALONG THE SEGMENTS.
C
      COMMON N, L, LEC, IMP
      DIMENSION X(51), Y(51), G(50, 50), FI(50), DFI(50)
      DIMENSION KODE (50), CX(50), CY(50), SOL(50), H(50, 50)
C
C  INITIALIZATION OF PROGRAM PARAMETERS
C  NX = MAXIMUM DIMENSION OF THE SISTEM OF EQUATIONS.
C
      NX = 50
C
C  ASSIGN DATA SET NUMBER FOR INPUT, LEC, AND OUTPUT, IMP
C
      LEC = 5
      IMP = 6
C
C  INPUT
C
      CALL INPUT (CX, CY, X, Y, KODE, FI)
C
C  FORM SYSTEM OF EQUATIONS
C
```

```
      CALL FMAT(X, Y, G, H, FI, DFI, KODE, NX)
C
C  SOLUTION OF THE SYSTEM OF EQUATIONS
C
      CALL SLNPD(G, DFI, D, N, NX)
C
C  COMPUTE THE POTENTIAL VALUES IN INTERNAL POINTS
C
      CALL INTER(FI, DFI, KODE, CX, CY, X, Y, SOL)
C
C  OUTPUT
C
      CALL OUTPUT(X, Y, FI, DFI, CX, CY, SOL)
      STOP
      END
```

Forming the system of equations – subroutine FMAT

The FMAT listing is now as follows:

```
      SUBROUTINE FMAT(X, Y, G, H, FI, DFI, KODE, NX)
C
C  PROGRAM 10
C
C  THIS SUBROUTINE COMPUTES THE G AND H MATRICES AND FORM THE
C  SYSTEM AX = F
C
      COMMON N, L, LEC, IMP
      DIMENSION X(1), Y(1), G(NX, NX), H(NX, NX), FI(1), KODE(1), DFI(1)
C
C  CLEAR G AND H MATRICES
C
      DO 10 I = 1, N
      DO 10 J = 1, N
      G(I, J) = 0
   10 H(I, J) = 0
C
C  COMPUTE G AND H MATRICES
C
      X(N + 1) = X(1)
      Y(N + 1) = Y(1)
      DO 110  I = 1, N
      NF = I + 1
      NS = I + N - 2
      DO 50  JJ = NF, NS
      IF(JJ - N)30, 30, 20
   20 J = JJ - N
      GO TO 40
   30 J = JJ
   40 CALL INTE(X(I), Y(I), X(J), Y(J), X(J + 1), Y(J + 1), A1, A2, B1, B2)
      IF (J - N)42, 43, 43
   42 H(I, J + 1) = H(I, J + 1) + A2
      G(I, J + 1) = G(I, J + 1) + B2
      GO TO 44
```

```
   43 H(I, 1) = H(I, 1) + A2
      G(I, 1) = G(I, 1) + B2
   44 H(I, J) = H(I, J) + A1
      G(I, J) = G(I, J) + B1
   50 H(I, I) = H(I, I) - A1 - A2
      NF = I + N - 1
      NS = I + N
      DO 95  JJ = NF, NS
      IF(JJ - N)70, 70, 60
   60 J = JJ - N
      GO TO 80
   70 J = JJ
   80 CALL INLO(X(J), Y(J), X(J + 1), Y(J + 1), B1, B2)
      IF(JJ - NF)82, 82, 83
   82 CH = B1
      B1 = B2
      B2 = CH
   83 IF(J - N)85, 90, 90
   85 G(I, J + 1) = G(I, J + 1) + B2
      GO TO 95
   90 G(I, 1) = G(I, 1) + B2
   95 G(I, J) = G(I, J) + B1
  110 CONTINUE
C
C  ARRANGE THE SYSTEM OF EQUATIONS READY TO BE SOLVED
C
      DO 150 J = 1, N
      IF (KODE(J)) 150, 150, 140
  140 DO 150 I = 1, N
      CH = G(I, J)
      G(I, J) = -H(I, J)
      H(I, J) = -CH
  150 CONTINUE
C
C  DFI ORIGINALLY CONTAINS THE INDEPENDENT COEFFICIENTS,
C  AFTER SOLUTION IT WILL CONTAIN THE VALUES OF THE SYSTEM
C  UNKNOWNS
C
      DO 160 I = 1, N
      DFI(I) = 0.
      DO 160 J = 1, N
      DFI(I) = DFI(I) + H(I, J)*FI(J)
  160 CONTINUE
      RETURN
      END
```

Subroutine INTE

It replaces subroutine INTE of Program 4, but instead of computing only one value for the element along which the integration is done, this subroutine now computes the part of the G and H coefficients corresponding to the adjacent nodes.

```
      SUBROUTINE INTE(XP, YP, X1, Y1, X2, Y2, A1, A2, B1, B2)
C
C  PROGRAM 11
C
C  THIS SUBROUTINE COMPUTES THE INTEGRALS ALONG AN ELEMENT
C  WHICH DOES NOT INCLUDE THE NODE UNDER CONSIDERATION
C
C  DIST = DISTANCE FROM THE POINT UNDER CONSIDERATION TO THE
C  BOUNDARY ELEMENTS
C  RA = DISTANCE FROM THE POINT UNDER CONSIDERATION TO THE
C  INTEGRATION POINTS IN THE BOUNDARY ELEMENTS
C
      DIMENSION XCO(4), YCO(4), GI(4), OME(4)
      GI(1) = 0.86113631
      GI(2) = -GI(1)
      GI(3) = 0.33998104
      GI(4) = -GI(3)
      OME(1) = 0.34785485
      OME(2) = OME(1)
      OME(3) = 0.65214515
      OME(4) = OME(3)
      AX = (X2 - X1)/2
      BX = (X2 + X1)/2
      AY = (Y2 - Y1)/2
      BY = (Y2 + Y1)/2
      IF(AX)10, 20, 10
   10 TA = AY/AX
      DIST = ABS((TA*XP - YP + Y1 - TA*X1)/SQRT(TA**2 + 1))
      GO TO 30
   20 DIST = ABS(XP - X1)
   30 SIG = (X1 - XP)*(Y2 - YP) - (X2 - XP)*(Y1 - YP)
      IF(SIG)31, 32, 32
   31 DIST = -DIST
   32 A1 = 0.
      A2 = 0.
      B1 = 0.
      B2 = 0.
      DO 40 I = 1, 4
      XCO(I) = AX*GI(I) + BX
      YCO(I) = AY*GI(I) + BY
      RA = SQRT((XP - XCO(I))**2 + (YP - YCO(I))**2)
      H = DIST*OME(I)*SQRT(AX**2 + AY**2)/RA**2
      G = ALOG(1/RA)*OME(I)*SQRT(AX**2 + AY**2)
      A1 = A1 + (GI(I) - 1)*H/2
      A2 = A2 - (GI(I) + 1)*H/2
      B1 = B1 - (GI(I) - 1)*G/2
   40 B2 = B2 + (GI(I) + 1)*G/2
      RETURN
      END
```

Subroutine INLO

It replaces the previous INLO of Program 5 and computes the part of the elements of the matrix G corresponding to the integrals along the elements which include the node under consideration.

```
      SUBROUTINE INLO(X1, Y1, X2, Y2, B1, B2)
C
C  PROGRAM 12
C
C  THIS SUBROUTINE COMPUTES THE INTEGRALS ALONG AN ELEMENT
C  INCLUDING THE NODE UNDER CONSIDERATION
C
      SEP = SQRT((X2 – X1)**2 + (Y2 – Y1)**2)
      B1 = SEP*(1.5 – ALOG(SEP))/2
      B2 = SEP*(0.5 – ALOG(SEP))/2
      RETURN
      END
```

Subroutine INTER

This routine replaces the INTER subroutine, Program 7. It puts all the potentials in FI and the derivatives in DFI and finally computes the values of the potential for the internal points. The potential at the internal points is again given by Equation (3.25) but the G_{ii} and $\hat{H}_{ij}$ coefficients are defined by Equations (3.35) to (3.37).

```
      SUBROUTINE INTER(FI, DFI, KODE, CX, CY, X, Y, SOL)
C
C  PROGRAM 13
C
C  THIS SUBROUTINE COMPUTES THE POTENTIAL VALUE AT INTERNAL
C  POINTS
C
      COMMON N, L, LEC, IMP
      DIMENSION FI(1), DFI(1), KODE(1), CX(1), CY(1), X(1), Y(1), SOL(1)
C
C  REORDER FI AND DFI ARRAY TO PUT ALL THE VALUES OF THE POTENTIAL
C  IN FI AND ALL THE VALUES OF THE DERIVATIVE IN DFI
C
      DO 20 I = 1, N
      IF (KODE(I)) 20, 20, 10
   10 CH = FI(I)
      FI(I) = DFI(I)
      DFI(I) = CH
   20 CONTINUE
C
C  COMPUTE THE POTENTIAL VALUES FOR INTERNAL POINTS
C
      DO 40  K = 1, L
      SOL(K) = 0.
      DO 30  J = 1, N
      CALL INTE(CX(K), CY(K), X(J), Y(J), X(J + 1), Y(J + 1), A1, A2, B1, B2)
      IF(J – N)32, 33, 33
   32 SOL(K) = SOL(K) + DFI(J)*B1 + DFI(J + 1)*B2 – FI(J)*A1  – FI(J + 1)*A2
      GO TO 30
   33 SOL(K) = SOL(K) + DFI(J)*B1 + DFI(1)*B2 – FI(J)*A1 – FI(1)*A2
   30 CONTINUE
   40 SOL(K) = SOL(K)/(2*3.1415926)
      RETURN
      END
```

Subroutine OUTPT

The OUTPT subroutine is like the one used in Program 8, but instead of printing the arrays XM and YM for the nodal coordinates it now prints the X and Y arrays.

```
      SUBROUTINE OUTPT(X, Y, FI, DFI, CX, CY, SOL)
C
C  PROGRAM 14
C
      COMMON N, L, LEC, IMP
      DIMENSION X(1), Y(1), FI(1), DFI(1), CX(1), CY(1), SOL(1)
      WRITE (IMP, 100)
100   FORMAT (' ', 120('*')//1X, 'RESULTS'//2X, 'BOUNDARY NODES'//16X,
     1 'X', 23X, 'Y', 19X, 'POTENTIAL', 10X, 'POTENTIAL DERIVATIVE'/)
      DO 10 I = 1, N
10    WRITE (IMP, 200) X(I), Y(I), FI(I), DFI(I)
200   FORMAT (4(10X, E14.7))
      WRITE (IMP, 300)
300   FORMAT(//, 2X, 'INTERNAL POINTS', //11X, 'X', 18X, 'Y', 14X,
     1 'POTENTIAL',/)
      DO 20 K = 1, L
20    WRITE (IMP, 400)CX(K), CY(K), SOL(K)
400   FORMAT (3(5X, E14.7))
      WRITE (IMP, 500)
500   FORMAT (' ', 120('*'))
      RETURN
      END
```

Example 3.1 [12]

The simple potential example shown in Figure 3.10 was solved to give some insight into the different types of elements and strategies that can be used. The boundary of the domain under consideration was divided into 12 elements. Figure 3.11(a) shows the subdivision for constant elements while in Figure 3.11(b) linear elements are used.

The constant elements solution proved to give reasonable agreement with the known trivial solution on the boundaries and inside the domain. It may appear that an improved solution will be obtained if linear elements are used on the boundary (Figure 3.11(b)). Linear elements are also attractive as they lend themselves to be joined with finite elements if necessary. Unfortunately problems will now appear at corner points, such as those shown in Figures 3.11(b), which can have two values for the normal derivative $\partial u/\partial n$ depending on the side under consideration. At these points one needs to select which of the two variables u or $\partial u/\partial n$ will be prescribed.

As $\partial u/\partial n$ cannot be uniquely defined one generally will choose to prescribe u. This however, produces a not very accurate computed value for the derivatives at the corners (Figure 3.11(b)). This problem does not

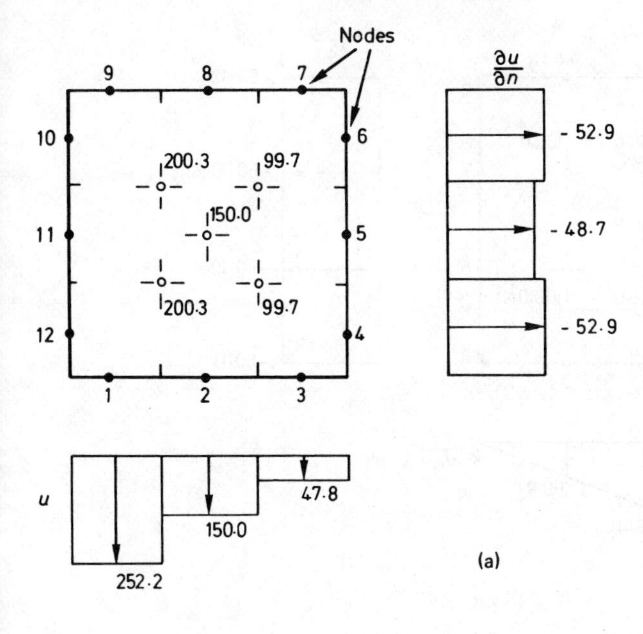

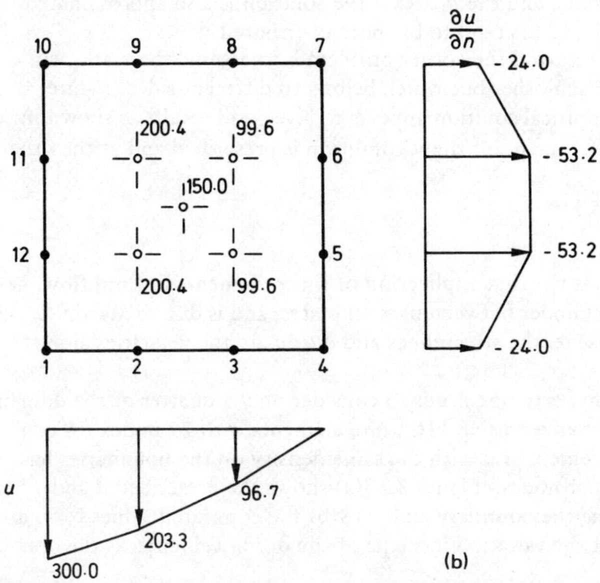

Fig. 3.11. Solution with (a) constant elements, (b) linear elements and (c) linear elements and two nodes at the corners. (Note that at corners the value of u *is defined but not that of* ∂u/∂n*)*

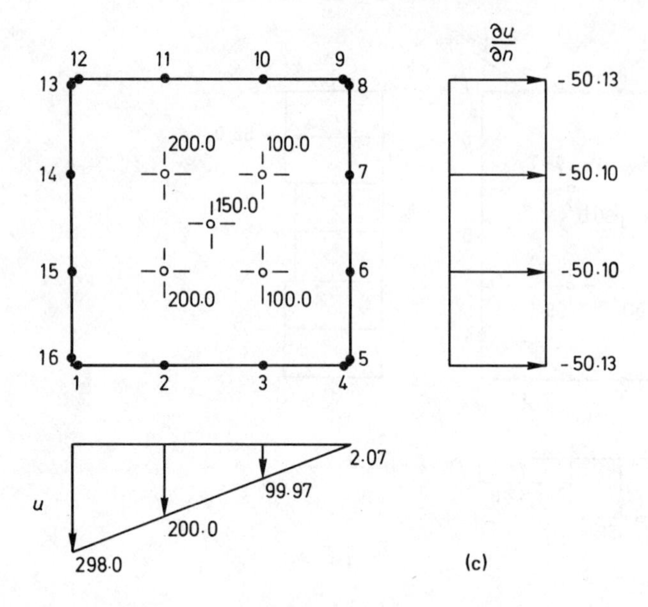

Fig. 3.11 (contd.)

occur in finite elements due to the way in which the natural boundary conditions are prescribed and the fact that the solution is also approximated in the domain, i.e., errors tend to be more distributed.

A simple way to avoid the corner problem is to assume there are two points very near each other but which belong to different sides (Figure 3.11(c)). This empirical solution appears to give good results as shown in Figure 3.11(c). At one point the u condition is prescribed and at the other the $\partial u/\partial n$.

Example 3.2

Figure 3.12 depicts the first application of finite elements to fluid flow, i.e. a flow around a cylinder between parallel plates, and is due to Martin[13]. Function u represents the streamlines and $\partial u/\partial n$ are the velocities along the boundaries.

Because of symmetry one needs to consider only a quarter of the domain. This has been discretized using 110 finite elements with 72 nodes. A constant boundary element grid with the same density on the boundaries has only 32 elements or nodes. Figure 3.13(a) shows the prescribed u and $\partial u/\partial n$ values along the boundary and 3.13(b) the computed values for u and $\partial u/\partial n$. Note that the variation in shape of the $\partial u/\partial n$ velocities is the one to be expected.

The use of linear elements in this case tends to give better results for the u and $\partial u/\partial n$ functions along the boundary but there may be a distortion of the

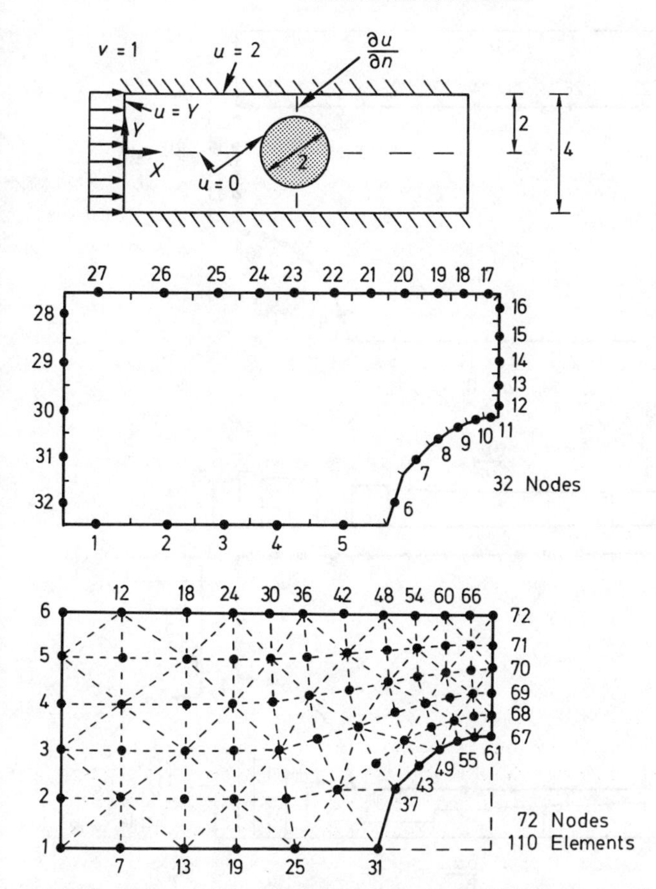

Fig.3.12. Flow around a cylinder between parallel plates. Constant boundary elements versus finite element grid

value of $\partial u/\partial n$ near those corners in which the normal derivative can take two very different values, depending on the side under consideration (Figure 3.14). To avoid this the $\partial u/\partial n$ are prescribed at points 32 and 37, for instance, while the value of the u function is given at 1 and 31. Values of u inside the domain agree well with the finite elements results. No exact solution is known for this case. The values of $\partial u/\partial n$ calculated on the boundaries using finite elements are unsatisfactory (Figure 3.15), while both boundary element solutions represent well the expected flow configuration.

3.6 QUADRATIC AND HIGHER ORDER ELEMENTS

Quadratic elements (or other higher order elements) can be used to better represent the geometry of the body. They do not present any special dif-

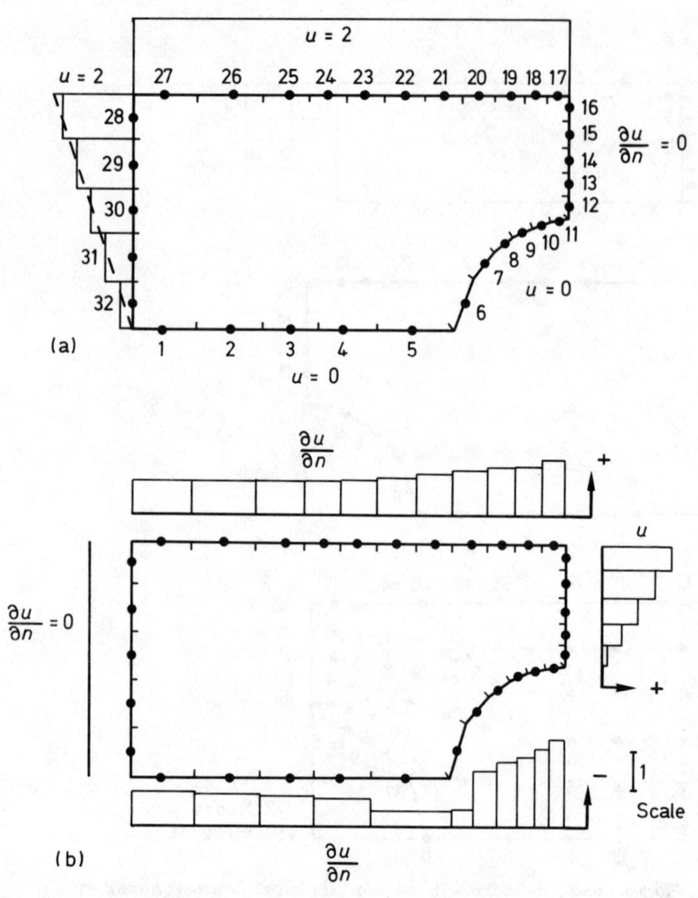

Fig.3.13. Prescribed conditions and solutions for constant elements. (a) Values of u and ∂u/∂n prescribed along the boundary. (b) Computed values for u and ∂u/∂n along the boundary

ficulty but a Jacobian transformation is required to pass from the Γ to the ξ system (Figure 3.16).

The starting point for the model is Equation (3.32) and the u and q functions can now be written as,

$$u(\xi) = \phi_1 u_1 + \phi_2 u_2 + \phi_3 u_3 = [\phi_1 \; \phi_2 \; \phi_3] \begin{Bmatrix} u_1 \\ u_2 \\ u_3 \end{Bmatrix}$$

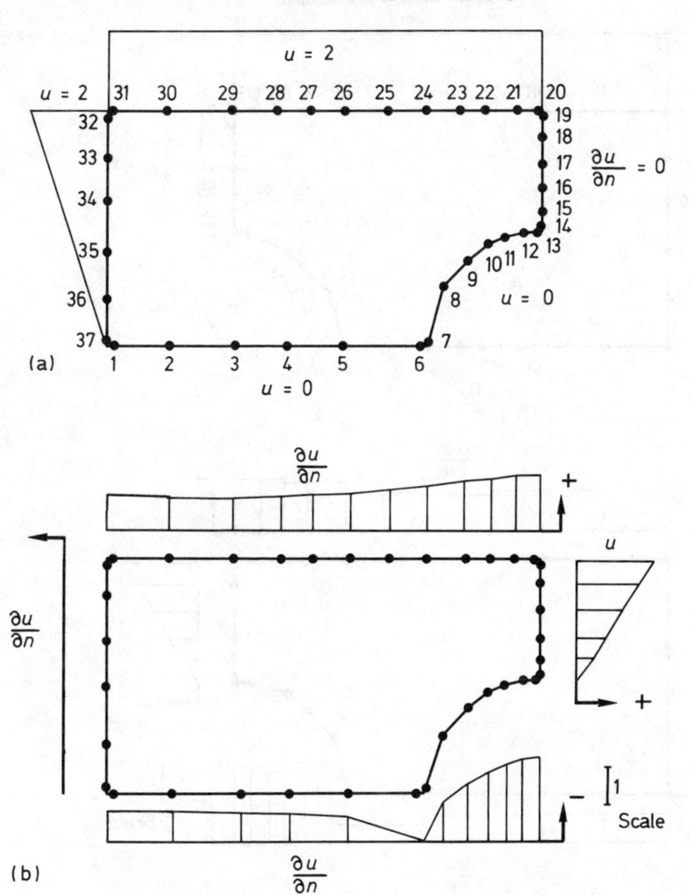

Fig.3.14. Prescribed conditions and solutions for linear elements. (a) Values for ∂u/∂n and prescribed along the boundary. (b) Computed values of u and ∂u/∂n along the boundary

$$q(\xi) = \phi_1 q_1 + \phi_2 q_2 + \phi_3 q_3 = [\phi_1\ \phi_2\ \phi_3] \begin{Bmatrix} q_1 \\ q_2 \\ q_3 \end{Bmatrix} \tag{3.42}$$

where

$$\phi_1 = \tfrac{1}{2}\xi(\xi - 1), \quad \phi_2 = \tfrac{1}{2}\xi(1 + \xi), \quad \phi_3 = (1 - \xi)(1 + \xi)$$

Note that these functions are such that they give the nodal value of the functions when specialized for the nodes (see table in Figure 3.17) and that

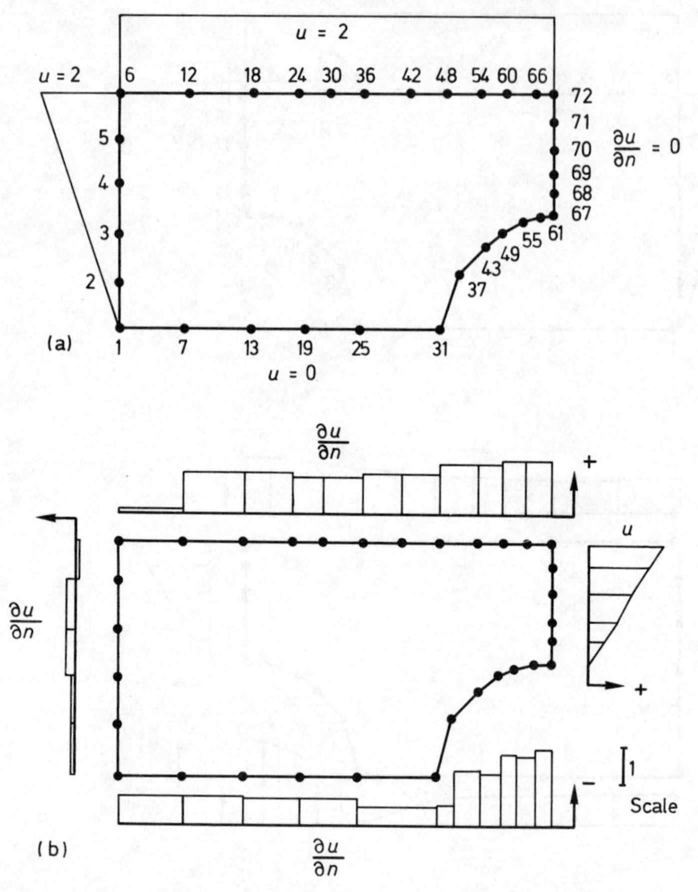

Fig.3.15. (a) Values of u and ∂u/∂n prescribed along the boundary. (b) Computed values for u and ∂u/∂n along the boundary.

they vary quadratically.

A typical integral, along an element 'j' is now,

$$\int_{\Gamma_j} u^* d\Gamma = [h_{j1} \; h_{j2} \; h_{j3}] \begin{Bmatrix} u_1 \\ u_2 \\ u_3 \end{Bmatrix} \tag{3.43}$$

Now, however, the evaluation of the integrals generally requires the use of a Jacobian, as ϕ_i are functions of ξ and the integrals are on Γ. For a two-

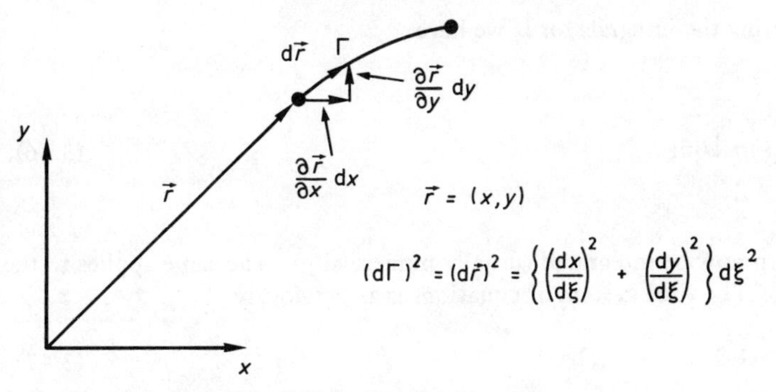

Fig.3.16. Geometrical definitions for curved boundary

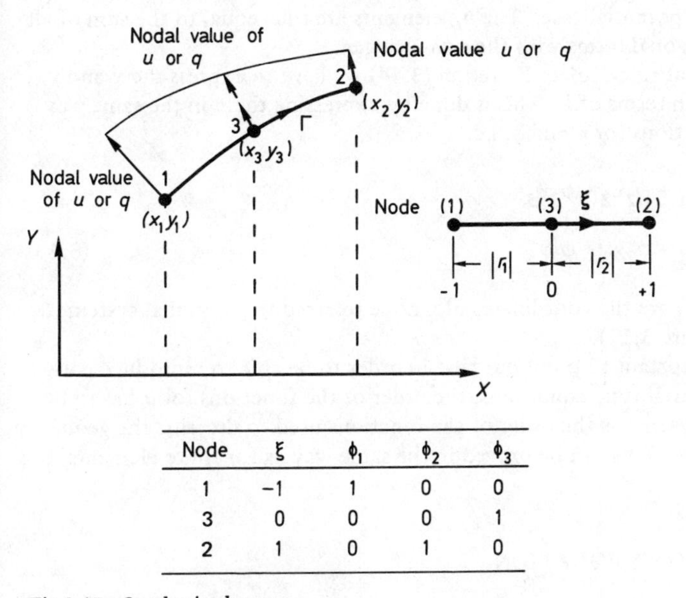

Node	ξ	ϕ_1	ϕ_2	ϕ_3
1	−1	1	0	0
3	0	0	0	1
2	1	0	1	0

Fig.3.17. Quadratic element

dimensional problem this is easily done as the Jacobian is simply (see Figure 3.16),

$$|J| = \frac{\partial\Gamma}{\partial\xi} = \left\{\left(\frac{\mathrm{d}x}{\mathrm{d}\xi}\right)^2 + \left(\frac{\mathrm{d}y}{\mathrm{d}\xi}\right)^2\right\}^{1/2} \tag{3.44}$$

Hence,

$$\mathrm{d}\Gamma = |J|\,\mathrm{d}\xi \tag{3.45}$$

Substituting the integrals for H we have

$$\int_{(1)}^{(2)} u(\xi) q^* |J| d\xi \qquad (3.46)$$

which can now be integrated (usually numerically). The same applies to the g_{ji} terms. The final system of equations is as previously

$$\mathbf{H}\,\mathbf{U} = \mathbf{G}\,\mathbf{Q} \qquad (3.47)$$

The same considerations about the diagonal terms in **H** apply here as for the linear and constant elements, i.e. that we can evaluate h_{ii} by assuming a constant potential case. The h_{ii} elements are thus equal to the sum of all the off-diagonal terms with the sign changed.

Note that to calculate Equation (3.44) we have to express the x and y functions in terms of ξ. This is done by expressing them in the same way as the functions for u and q, i.e.

$$x = \phi_1 x_1 + \phi_2 x_2 + \phi_3 x_3$$

$$y = \phi_1 y_1 + \phi_2 y_2 + \phi_3 y_3 \qquad (3.48)$$

where x_i, y_i are the coordinates of a node referred to the global system X, Y (see Figure 3.17).

It is important to point out that in order to be able to reproduce constant potential type conditions, the order of the functions for u has to be at least the same as the order of the functions used to describe the geometry of the body. This can be proved in the same way as for finite elements[14].

3.7 POISSON'S EQUATION

Consider now the case of a function u which is required to satisfy the following Poisson equation in Ω,

$$\nabla^2 u - p = 0 \qquad (3.49)$$

As previously, the boundary conditions corresponding to this problem are:

Essential conditions such as $u = \bar{u}$ on Γ_1.
Natural conditions $\partial u/\partial n = \bar{q}$ on Γ_2.

The total boundary is $\Gamma = \Gamma_1 + \Gamma_2$.

The weighting function u^* now produces the following statement

$$\int_{\Omega} (\nabla^2 u - p)u^* \mathrm{d}\Omega = \int_{\Gamma_2} \left(\frac{\partial u}{\partial n} - \bar{q} \right) u^* \mathrm{d}\Gamma - \int_{\Gamma_1} (u - \bar{u}) \frac{\partial u^*}{\partial n} \mathrm{d}\Gamma \quad (3.50)$$

Integrating the first term on the left hand side by parts twice we obtain,

$$- \int_{\Omega} pu^* \mathrm{d}\Omega + \int_{\Omega} u \nabla^2 u^* \mathrm{d}\Omega = - \int_{\Gamma_2} \bar{q} u^* \mathrm{d}\Omega - \int_{\Gamma_1} \frac{\partial u}{\partial n} u^* \mathrm{d}\Gamma$$

$$+ \int_{\Gamma_2} u \frac{\partial u^*}{\partial n} \mathrm{d}\Gamma + \int_{\Gamma_1} \bar{u} \frac{\partial u^*}{\partial n} \mathrm{d}\Gamma \quad (3.51)$$

We can use the same fundamental solutions for u^* as the ones previously used in Equation (3.5), i.e.

$$\nabla^2 u^* + \Delta^i = 0 \quad (3.52)$$

Hence Equation (3.51) at a point 'i' becomes,

$$\int pu^* \mathrm{d}\Omega + u^i + \int_{\Gamma_2} uq^* \mathrm{d}\Gamma + \int_{\Gamma_1} \bar{u} q^* \mathrm{d}\Gamma = \int_{\Gamma_2} \bar{q} u^* \mathrm{d}\Gamma + \int_{\Gamma_1} qu^* \mathrm{d}\Gamma \quad (3.53)$$

As we take 'i' to the boundary the u^i term in Equation (3.53) will be multiplied by ½ for a smooth boundary. If the boundary is not smooth at the point 'i' instead of ½ we have a constant which can be determined from constant potential considerations and Equation (3.53) can be written as,

$$c^i u^i + \int_{\Omega} pu^* \mathrm{d}\Omega + \int_{\Gamma} uq^* \mathrm{d}\Gamma = \int_{\Gamma} qu^* \mathrm{d}\Gamma \quad (3.54)$$

Discretization using constant elements produces

$$c^i u^i + \int_{\Omega} pu^* \mathrm{d}\Omega + \sum_{j=1}^{n} u_j \int_{\Gamma_j} q^* \mathrm{d}\Gamma = \sum_{j=1}^{n} q_j \int_{\Gamma_j} u^* \mathrm{d}\Gamma \quad (3.55)$$

which can be written as,

$$c^i u^i + B_i + \sum_{j=1}^{n} \hat{H}_{ij} u_j = \sum_{j=1}^{n} G_{ij} q_j \quad (3.56)$$

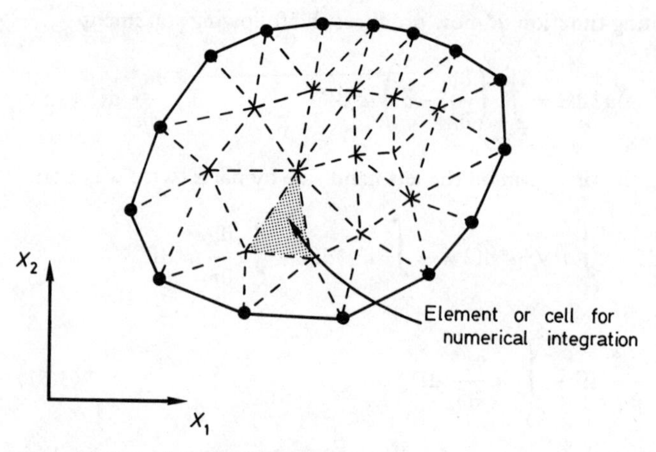

Fig.3.18. Boundary and internal elements

or

$$B_i + \sum_{j=1}^{n} H_{ij} u_j = \sum_{j=1}^{n} G_{ij} q_j$$

The B_i term is the result of having integrated the term

$$\int_{\Omega} p u^* \mathrm{d}\Omega.$$

The integration can be performed by dividing the domain Ω into a series of cells or 'elements' similar in shape to those used in the finite element method (see Figure 3.18), but conceptually different. Over each cell a numerical integration formula can be applied, hence

$$B_i = \sum_{n_e} \left(\sum_{j=1}^{k} w_j (pu^*)_j \right) A_{n_e} \tag{3.57}$$

where n_e = number of elements; k = number of integration points on each element; w_j = weighting function and A_{n_e} = area of cell or element.

The whole set of equations for the n nodes can be expressed in matrix form as

$$\mathbf{B} + \mathbf{H}\,\mathbf{U} = \mathbf{G}\,\mathbf{Q} \tag{3.58}$$

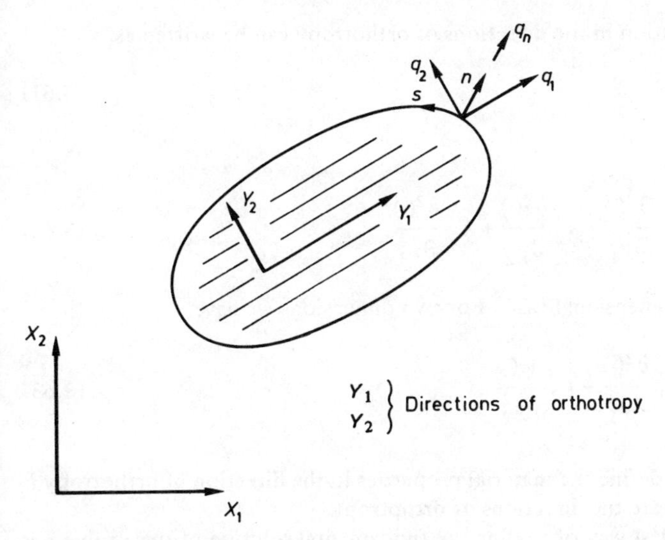

Fig. 3.19. Orthotropic body

Note that n_1 values of u and n_2 values of q are known on the boundary. Hence Equations (3.58) can be reordered in such a way that all the unknowns are on the left hand side, i.e.

$$\mathbf{A}\,\mathbf{X} = \mathbf{F} \tag{3.59}$$

where $\mathbf{X}$ is the vector of unknowns u's and q's.

Once the values of u and q on the whole boundary are known one can calculate the u at any interior point, i.e.

$$u^i = \sum_{j=1}^{n} G_{ij}q_j - \sum_{j=1}^{n} \hat{H}_{ij}u_j - B_i \tag{3.60}$$

3.8 ORTHOTROPIC CASE

In many engineering applications the material properties cannot be considered to be isotropic and we need to consider them as orthotropic. Consider the case of a body such as the one shown in Figure 3.19. The equilibrium equation in the directions of orthotropy can be written as,

$$\nabla_0^2 u = 0 \tag{3.61}$$

where

$$\nabla_0^2(\,) = k_1 \frac{\partial^2(\,)}{\partial y_1^2} + k_2 \frac{\partial^2(\,)}{\partial y_2^2} + k_3 \frac{\partial^2(\,)}{\partial y_3^2} \tag{3.62}$$

for a three-dimensional case. For two dimensions we have,

$$\nabla_0^2(\,) = k_1 \frac{\partial^2(\,)}{\partial y_1^2} + k_2 \frac{\partial^2(\,)}{\partial y_2^2} \tag{3.63}$$

the k_i terms define the material properties in the direction of orthotropy i. Note that y_i are the directions of orthotropy.

The simplest way of finding the fundamental solution of this problem is by using the following transformation:

$$z_i = \frac{y_i}{\sqrt{k_i}} \tag{3.64}$$

If we assume that a concentrated potential is acting at a point 'i', the fundamental solution should satisfy

$$\nabla_0^2 u^* + \Delta^i = 0 \tag{3.65}$$

where Δ^i can be written as $\Delta(y_1 - y_1^i)\Delta(y_2 - y_2^i)\Delta(y_3 - y_3^i)$ for three dimensions. This is another way of saying that the integral of Δ^i will be 1 at the point 'i'. The reason we have written it this way is to indicate that Δ is also a function of the coordinates.

Substituting Equation (3.64) into (3.65), taking into account that Δ^i is function of the coordinates produces

$$\nabla^2 u^* + \Delta^i = \frac{\partial^2 u^*}{\partial z_1^2} + \frac{\partial^2 u^*}{\partial z_2^2} + \frac{\partial^2 u^*}{\partial z_3^2} + \Delta^i = 0 \tag{3.66}$$

where Δ^i is now

$$\Delta^i = \Delta(\sqrt{k_1}(z_1 - z_1^i))\Delta(\sqrt{k_2}(z_2 - z_2^i))\Delta(\sqrt{k_3}(z_3 - z_3^i))$$

Equation (3.66) has now to be integrated, i.e.

$$\int_{\Omega} u(\nabla^2 u^* + \Delta^i)\mathrm{d}\Omega = \int_{\Omega} u\nabla^2 u^* \mathrm{d}\Omega + \int_{\Omega} u\Delta^i \mathrm{d}\Omega \tag{3.67}$$

The first term in Equation (3.67) is satisfied by the fundamental solution. The second can now be rewritten, taking into account that the integral of the Dirac delta function is

$$\lim \int_{-\epsilon+z^i}^{+\epsilon+z^i} u(z)\Delta(\sqrt{k}(z - z^i))\mathrm{d}z = \lim \frac{1}{\sqrt{k}} \int_{\xi^i-(\epsilon/\sqrt{k})}^{\xi^i+(\epsilon/\sqrt{k})} u\left(\frac{\xi}{\sqrt{k}}\right) \Delta(\xi - \xi^i)\mathrm{d}\xi$$

$$= \frac{1}{\sqrt{k}}\, u\left(\frac{\xi}{\sqrt{k}}\right) = \frac{1}{\sqrt{k}}\, u(z)$$

where $\xi = \sqrt{k}z$ is a variable used for the integration. Hence the second term on the right hand side of Equation (3.67) becomes,

$$\int u\Delta^i \mathrm{d}\Omega = \frac{1}{\sqrt{k_1 k_2 k_3}}\, u^i \tag{3.68}$$

The fundamental solution for Equation (3.66) is the same as for (3.8) i.e.

$$u^* = \frac{1}{4\pi r} \tag{3.69}$$

where

$$r = (z_1^2 + z_2^2 + z_3^2)^{1/2} = \left(\frac{y_1^2}{k_1} + \frac{y_2^2}{k_2} + \frac{y_3^2}{k_3}\right)^{1/2} \tag{3.70}$$

In order for the value of the integral in Equation (3.68) to be equal to u^i it is customary to take

$$u^* = \frac{1}{(k_1 k_2 k_3)^{1/2}}\, \frac{1}{4\pi}\, \frac{1}{\left(\frac{y_1^2}{k_1} + \frac{y_2^2}{k_2} + \frac{y_3^2}{k_3}\right)^{1/2}} \tag{3.71}$$

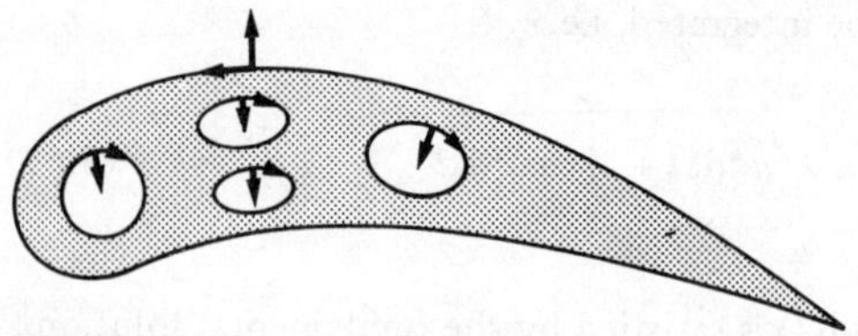

Fig. 3.20. Body with holes

The same transformation can be applied to two dimensional problems to produce,

$$u^* = \frac{1}{(k_1 k_2)^{1/2}} \frac{1}{2\pi} \ln \left\{ \frac{1}{\left(\dfrac{y_1^2}{k_1} + \dfrac{y_2^2}{k_2} \right)^{1/2}} \right\} \tag{3.72}$$

Flux at the boundary

Consider the two dimensional case for simplicity. To calculate the fluxes at the boundary one can apply Green's theorem, i.e.

$$\int_\Omega \left(k_1 \frac{\partial^2 u}{\partial y_1^2} + k_2 \frac{\partial^2 u}{\partial y_2^2} \right) \mathrm{d}\Omega \Rightarrow \int_\Gamma \left(k_1 \frac{\partial u}{\partial y_1} n_1 + k_2 \frac{\partial u}{\partial y_2} n_2 \right) \mathrm{d}\Gamma \tag{3.73}$$

The term between brackets in the r.h.s. integral is the internal flux. Hence

$$\bar{q} = q = k_1 \frac{\partial u}{\partial y_1} n_1 + k_2 \frac{\partial u}{\partial y_2} n_2 \tag{3.74}$$

where $n_1 = \cos(n, y_1)$ and $n_2 = \cos(n, y_2)$. This expression will allow us to calculate the values of q^* needed for the boundary element solution.

3.9 PROBLEMS WITH MORE THAN ONE SURFACE

The method can easily be used to study problems with more than one surface, such as the case of a body with holes illustrated in Figure 3.20. In order to define an external or internal boundary we need to identify the direction of the normal. This can be done for two dimensional problems by adopting the following rule.

(1) For external surfaces the numbering scheme is defined in the counterclockwise direction.

(2) For internal surfaces the numbering scheme is defined in the clockwise direction.

With these rules the normal can be well defined in the computer program.

The following computer program is based in the constant element program previously defined but can take up to five different surfaces. Each surface is numbered according to the above rule. The subroutines are as follows.

Main program

This program is the same as previously (Program 1) except that the COMMON is replaced by

COMMON N, L, NC(5), M, LEC, IMP

where the new variables NC and M are

M: Number of difference surfaces
NC: Number of the last node of each different surface.

Subroutine INPUT

The input required is the same as previously (Program 2) with the exception of M and NC which are read in the same card as N and L (number of elements and number of internal points where the function is calculated). The listing is given below for completeness.

```
      SUBROUTINE INPUT(CX, CY, X, Y, KODE, FI)
C
C  PROGRAM 15
C
      COMMON N, L, NC(5), M, LEC, IMP
      DIMENSION CX(1), CY(1), X(1), Y(1), KODE(1), FI(1), TITLE (18)
C
C  N = NUMBER OF BOUNDARY ELEMENTS
C  L = NUMBER OF INTERNAL POINTS WHERE THE FUNCTION IS CALCULATED
C
      WRITE (IMP, 100)
  100 FORMAT(' ', 120('*'))
C
C  READ NAME OF THE JOB
C
      READ (LEC, 150) TITLE
  150 FORMAT (18A4)
      WRITE (IMP, 250) TITLE
  250 FORMAT (25X, 18A4)
```

```
C
C  READ BASIC PARAMETERS
C
        READ (LEC, 200) N, L, M, (NC(K), K = 1, 5)
  200   FORMAT (8I5)
        WRITE (IMP, 300) N, L
  300   FORMAT (//'DATA'//2X, 'NUMBER OF BOUNDARY ELEMENTS =',
      1 I3/1X, 'NUMBER OF INTERNAL POINTS WHERE THE FUNCTION IS
      2 CALCULATED =', I3)
        IF (M) 40, 40, 30
   30   WRITE (IMP, 999)M, (NC(K), K = 1, M)
  999   FORMAT(/2X, 'NUMBER OF DIFFERENT SURFACES = ', I3/2X, 'LAST
      1 NODES IN THESE SURFACES', 5(2X, I3))
C
C   READ INTERNAL POINTS COORDINATES
C
   40   DO 1 I = 1, L
    1   READ (LEC, 400) CX(I), CY(I)
  400   FORMAT (2F10.4)
C
C   READ COORDINATES OF EXTREME POINTS OF THE BOUNDARY
C   ELEMENTS IN ARRAY X AND Y
C
        WRITE (IMP, 500)
  500   FORMAT (//2X, 'COORDINATES OF THE EXTREME POINTS OF THE
      1 BOUNDARY ELEMENTS', //4X, 'POINT', 10X, 'X', 18X, 'Y')
        DO 10 I = 1, N
        READ (LEC, 600) X(I), Y(I)
  600   FORMAT (2F10.4)
   10   WRITE (IMP, 700) I, X(I), Y(I)
  700   FORMAT (5X, I3, 2(5X, E14.7))
C
C   READ BOUNDARY CONDITIONS
C   FI(I) = VALUE OF THE POTENTIAL OF THE NODE I IF KODE = 0.
C   VALUE OF THE POTENTIAL DERIVATIVE IF KODE = 1.
C
        WRITE (IMP, 800)
  800   FORMAT(//2X, 'BOUNDARY CONDITIONS'//5X, 'NODE', 6X, 'CODE',
      1 5X, 'PRESCRIBED VALUE')
        DO 20 I = 1, N
        READ (LEC, 900) KODE (I), FI(I)
  900   FORMAT (I5, F10.4)
   20   WRITE (IMP, 950) I, KODE(I), FI(I)
  950   FORMAT (5X, I3, 8X, I1, 8X, E14.7)
        RETURN
        END
```

Subroutine FMAT

The COMMON has changed and in addition there are some extra commands to differentiate each of the surfaces, which are needed to calculate the mid-point coordinates XM, YM. Note that the last nodal coordinate of each surface is in the mid point of the segment defined from the last extreme point to the first point on that surface.

The following program replaces Program 3:

```
      SUBROUTINE FMAT(X, Y, XM, YM, G, H, FI, DFI, KODE, NX)
C
C     PROGRAM 16
C
C     THIS SUBROUTINE COMPUTES G AND H MATRICES AND FORM THE
C     SYSTEM AX = F
C
      COMMON N, L, NC(5), M, LEC, IMP
      DIMENSION X(1), Y(1), XM(1), YM(1), G(NX, NX), H(NX, NX), F(1), KODE(1)
      DIMENSION DFI(1)
C
C     COMPUTE THE MID-POINT COORDINATES AND STORE IN ARRAY XM
C     AND YM
C
      X(N + 1) = X(1)
      Y(N + 1) = Y(1)
      DO 10 I = 1, N
      XM(I) = (X(I) + X(I + 1))/2
   10 YM(I) = (Y(I) + Y(I + 1))/2
      IF(N - 1)15, 15, 12
   12 XM(NC(1)) = (X(NC(1)) + X(1))/2
      YM(NC(1)) = (Y(NC(1)) + Y(1))/2
      DO 13 K = 2, M
      XM(NC(K)) = (X(NC(K)) + X(NC(K - 1) + 1))/2
   13 YM(NC(K)) = (Y(NC(K)) + Y(NC(K - 1) + 1))/2
C
C     COMPUTE G AND H MATRICES
C
   15 DO 30 I = 1, N
      DO 30 J = 1, N
      IF(M - 1)16, 16, 17
   17 IF(J - NC(1))19, 18, 19
   18 KK = 1
      GO TO 23
   19 DO 22 K = 2, M
      IF(J - NC(K))22, 21, 22
   21 KK = NC(K - 1) + 1
      GO TO 23
   22 CONTINUE
   16 KK = J + 1
   23 IF(I - J)20, 25, 20
   20 CALL INTE(XM(I), YM(I), X(J), Y(J), X(KK), Y(KK), H(I, J), G(I, J))
      GO TO 30
   25 CALL INLO(X(J), Y(J), X(KK), Y(KK), G(I, J))
      H(I, J) = 3.1415926
   30 CONTINUE
C
C     ARRANGE THE SYSTEM OF EQUATIONS READY TO BE SOLVED
C
      DO 50 J = 1, N
      IF(KODE(J))50, 50, 40
   40 DO 50 I = 1, N
      CH = G(I, J)
```

```
      G(I, J) = -H(I, J)
      H(I, J) = -CH
   50 CONTINUE
C
C  DFI ORIGINALLY CONTAINS THE INDEPENDENT COEFFICIENTS, AFTER
C  SOLUTION IT WILL CONTAIN THE VALUES OF THE SYSTEM UNKNOWNS
C
      DO 60 I = 1, N
      DFI(I) = 0.
      DO 60 J = 1, N
      DFI(I) = DFI(I) + H(I, J)*FI(J)
   60 CONTINUE
      RETURN
      END
```

Subroutine INTE

Same as Program 4.

Subroutine INLO

Same as Program 5.

Subroutine SLNPD

Same as Program 6.

Subroutine INTER

This subroutine varies from Program 7 to take into account the different surfaces.

```
      SUBROUTINE INTER(FI, DFI, KODE, CX, CY, X, Y, SOL)
C
C  PROGRAM 17
C
C  THIS SUBROUTINE COMPUTES THE POTENTIAL VALUE FOR INTERNAL
C  POINTS.
C
      COMMON N, L, NC(5), M, LEC, IMP
      DIMENSION FI(1), DFI(1), KODE(1), CX(1), CY(1), X(1), Y(1), SOL(1)
C
C  REORDER FI AND DFI ARRAY TO PUT ALL THE VALUES OF THE
C  POTENTIAL IN FI AND ALL THE VALUES OF THE DERIVATIVE IN DFI
C
```

```
      DO 20 I = 1, N
      IF(KODE(I))20, 20, 10
   10 CH = FI(I)
      FI(I) = DFI(I)
      DFI(I) = CH
   20 CONTINUE
C
C  COMPUTE THE POTENTIAL VALUES FOR INTERNAL POINTS
C
      DO 40 K = 1, L
      SOL(K) = 0.
      DO 30  J = 1, N
      IF(M - 1)28, 28, 22
   22 IF(J - NC(1))24, 23, 24
   23 KK = 1
      GO TO 29
   24 DO 26 LK = 2, M
      IF(J - NC(LK))26, 25, 26
   25 KK = NC(LK - 1) + 1
      GO TO 29
   26 CONTINUE
   28 KK = J + 1
   29 CALL INTE(CX(K), CY(K), X(J), Y(J), X(KK), Y(KK), A, B)
   30 SOL(K) = SOL(K) + DFI(J)*B - FI(J)*A
   40 SOL(K) = SOL(K)/(2*3.1415926)
      RETURN
      END
```

Subroutine OUTPUT

Same as Program 8 but with new COMMON, i.e.

COMMON N, L, NC(5), M, LEC, IMP

Example 3.3

As an example we solve a two-dimensional case (Figure 3.21) of a octahedron with a concentric hole. The number of elements is 16 and the number of different surfaces is 2. The program requires the last numbers of the segment nodes of each of the surfaces. They are 8 and 16. The boundary conditions are assumed to be zero potential on the external surface and 100 on the internal. Results at 4 internal points are also computed.

The data and results are shown in the following output.

```
************************************************************************
TEST OF THE COBE PROGRAM, HEAT FLOW BETWEEN TWO CIRCLES

DATA

NUMBER OF BOUNDARY ELEMENTS = 16
NUMBER OF INTERNAL POINTS WHERE THE FUNCTION IS CALCULATED = 4

NUMBER OF DIFFERENT SURFACES = 2
LAST NODES IN THESE SURFACES:  8   16
```

COORDINATES OF THE EXTREME POINTS OF THE BOUNDARY ELEMENTS

POINT	X	Y
1	0.0000000E 00	−0.2000000E 01
2	0.1414200E 01	−0.1414200E 01
3	0.2000000E 01	0.0000000E 00
4	0.1414200E 01	0.1414200E 01
5	0.0000000E 00	0.2000000E 01
6	−0.1414200E 01	0.1414200E 01
7	−0.2000000E 01	0.0000000E 00
8	−0.1414200E 01	−0.1414200E 01
9	−0.7071000E 00	−0.7071000E 00
10	−0.1000000E 01	0.0000000E 00
11	−0.7071000E 00	0.7071000E 00
12	0.0000000E 00	0.1000000E 01
13	0.7071000E 00	0.7071000E 00
14	0.1000000E 01	0.0000000E 00
15	0.7071000E 00	−0.7071000E 00
16	0.0000000E 00	−0.1000000E 01

BOUNDARY CONDITIONS

NODE	CODE	PRESCRIBED VALUE
1	0	0.0000000E 00
2	0	0.0000000E 00
3	0	0.0000000E 00
4	0	0.0000000E 00
5	0	0.0000000E 00
6	0	0.0000000E 00
7	0	0.0000000E 00
8	0	0.0000000E 00
9	0	0.1000000E 03
10	0	0.1000000E 03
11	0	0.1000000E 03
12	0	0.1000000E 03
13	0	0.1000000E 03
14	0	0.1000000E 03
15	0	0.1000000E 03
16	0	0.1000000E 03

RESULTS

BOUNDARY NODES

X	Y	POTENTIAL	POTENTIAL DERIVATI
0.7071000E 00	−0.1707100E 01	0.0000000E 00	−0.7293856E 02
0.1707100E 01	−0.7071000E 00	0.0000000E 00	−0.7293856E 02
0.1707100E 01	0.7071000E 00	0.0000000E 00	−0.7293856E 02
0.7071000E 00	0.1707100E 01	0.0000000E 00	−0.7293856E 02
−0.7071000E 00	0.1707100E 01	0.0000000E 00	−0.7293856E 02
−0.1707100E 01	0.7071000E 00	0.0000000E 00	−0.7293856E 02
−0.1707100E 01	−0.7071000E 00	0.0000000E 00	−0.7293856E 02
−0.7071000E 00	−0.1707100E 01	0.0000000E 00	−0.7293856E 02
−0.8535500E 00	−0.3535500E 00	0.1000000E 03	0.1502989E 03
−0.8535500E 00	0.3535500E 00	0.1000000E 03	0.1502989E 03
−0.3535500E 00	0.8535500E 00	0.1000000E 03	0.1502989E 03

0.3535500E 00	0.8535500E 00	0.1000000E 03	0.1503989E 03
0.8535500E 00	0.3535500E 00	0.1000000E 03	0.1502989E 03
0.8535500E 00	−0.3535500E 00	0.1000000E 03	0.1502989E 03
0.3535500E 00	−0.8535500E 00	0.1000000E 03	0.1502989E 03
−0.3535500E 00	−0.8535500E 00	0.1000000E 03	0.1502989E 03

INTERNAL POINTS

X	Y	POTENTIAL
0.1385820E 01	−0.5740250E 00	0.3156777E 02
0.1385820E 01	0.5740250E 00	0.3156777E 02
0.5740250E 00	0.1385820E 01	0.3156777E 02
−0.1385820E 01	−0.5740250E 00	0.3156777E 02

3.10 NON-HOMOGENEOUS SOLIDS

If a body is non-homogeneous we divide it into a series of regions each having the same properties (Figure 3.22). The regions can then be added together utilizing continuity of fluxes (equilibrium) and equalizing the potentials (continuity) at the interfaces.

Consider for simplicity a body consisting of only two regions as shown in Figure 3.23. The interface surface will be called Γ_I and we can define, on Part 1 of the domain the following functions,

$\mathbf{Q}^1$: Fluxes on external surface of Region 1.
$\mathbf{Q}_I^1$: Fluxes on interface Γ_I (considering it belongs to Region 1).

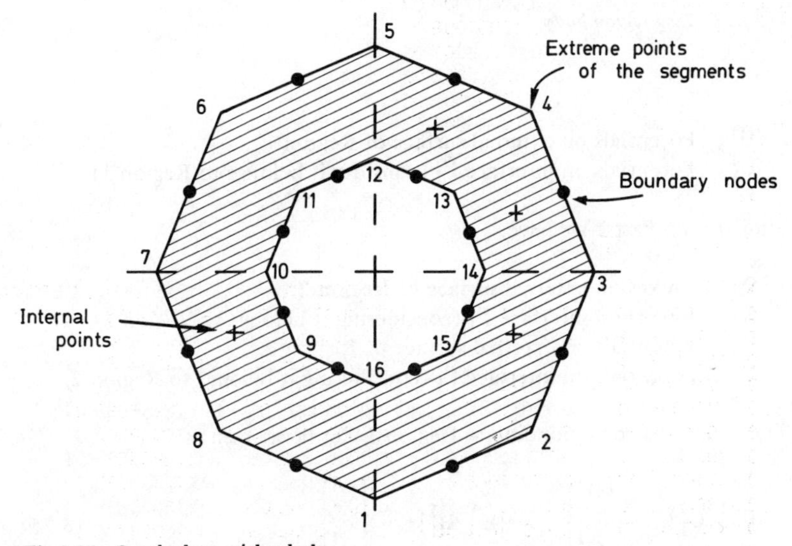

Fig.3.21. Octahedron with a hole

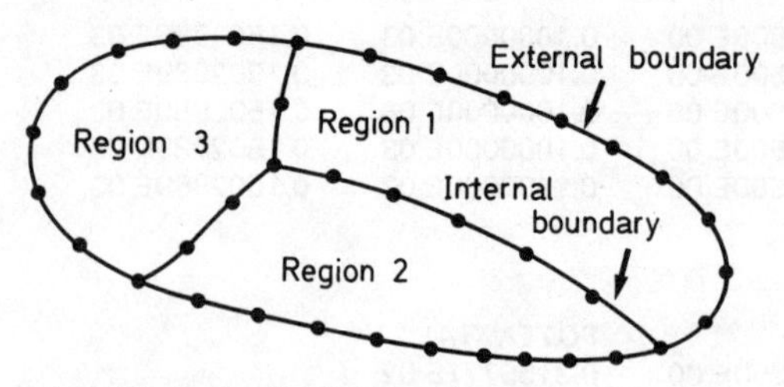

Fig.3.22. Body divided into elements

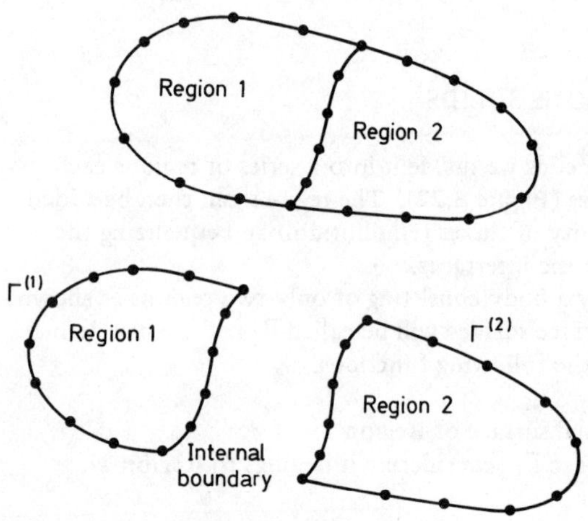

Fig.3.23. Two-region body

$\mathbf{U}^1$: Potentials on external surface of Region 1.
$\mathbf{U}_I^1$: Potentials on interface Γ_I (consider it belongs to Region 1).

Similarly on Part 2 we have,

$\mathbf{Q}^2$: Fluxes on external surface of Region 2.
$\mathbf{Q}_I^2$: Fluxes on interface Γ_I (considering it belongs to Region 2).
$\mathbf{U}^2$: Potentials on external surface of Region 2.
$\mathbf{U}_I^2$: Potentials on interface Γ_I (considering it belongs to Region 2).

The equations corresponding to Region 1 can be written as

$$[\mathbf{G}^1 \quad \mathbf{G}_I^1] \begin{Bmatrix} \mathbf{Q}^1 \\ \mathbf{Q}_I^1 \end{Bmatrix} = [\mathbf{H}^1 \quad \mathbf{H}_I^1] \begin{Bmatrix} \mathbf{U}^1 \\ \mathbf{U}_I^1 \end{Bmatrix} \tag{3.75}$$

and similarly for Region 2 we have that

$$[G^2 \; G_I^2] \begin{Bmatrix} Q^2 \\ Q_I^2 \end{Bmatrix} = [H^2 \; H_I^2] \begin{Bmatrix} U^2 \\ U_I^2 \end{Bmatrix} \tag{3.76}$$

Q_I and U_I indicate the fluxes and potentials at the interface Γ_I, where the following conditions apply;

(1) Compatibility: $U_I^1 = U_I^2 = U_I$ (3.77)

(2) Equilibrium: $Q_I^1 = -Q_I^2 = Q_I$ (3.78)

Hence Equation (3.75) and (3.76) can be written in function of U_I and Q_I as follows:

$$[G^1 \; G_I^1 \; -H_I^1] \begin{Bmatrix} Q^1 \\ Q_I \\ U_I \end{Bmatrix} = [H^1] \; \{U^1\} \tag{3.79}$$

$$[-G_I^2 \; -H_I^2 \; G^2] \begin{Bmatrix} Q_I \\ U_I \\ Q^2 \end{Bmatrix} = [H^2] \; \{U^2\} \tag{3.80}$$

These equations can be written as,

$$\begin{bmatrix} G^1 & G_I^1 & -H_I^1 & 0 \\ 0 & -G_I^2 & -H_I^2 & G^2 \end{bmatrix} \begin{Bmatrix} Q^1 \\ Q_I \\ U_I \\ Q^2 \end{Bmatrix} = \begin{bmatrix} H^1 & 0 \\ 0 & H^2 \end{bmatrix} \begin{Bmatrix} U^1 \\ U^2 \end{Bmatrix} \tag{3.81}$$

(Region 1 | Interface | Region 2)

If instead of all potentials prescribed we have a mixed problem we will need to reorder the system of equations in order to have all the unknowns on the left hand side and the known values on the right.

It is necessary to divide the body into regions when the material is non-homogeneous and also when the dimensions of the body are not very regular. For instance we may need to subdivide a long body into regions to avoid numerical inaccuracies.

3.11 THE HELMHOLTZ EQUATION

Another interesting potential type problem that can be solved using boundary elements is the case of the Helmholtz equation, which has numerous applications in engineering. The equation for the fundamental solution in two dimensions is

$$\frac{\partial^2 u^*}{\partial x^2} + \frac{\partial^2 u^*}{\partial y^2} + \omega^2 u^* + \Delta^i = 0 \tag{3.82}$$

for three-dimensional applications

$$\frac{\partial^2 u^*}{\partial x^2} + \frac{\partial^2 u^*}{\partial y^2} + \frac{\partial^2 u^*}{\partial z^2} + \omega^2 u^* + \Delta^i = 0 \tag{3.83}$$

It can be shown that the fundamental solution for the three-dimensional case is

$$u^* = -\frac{1}{4\pi r}\exp(-i\omega r) \tag{3.84}$$

and for two dimensions one obtains,

$$u^* = -\frac{i}{4}H_0^{(2)}(\omega r) \tag{3.85}$$

where $H_0^{(2)}$ is a Hankel function of zero order and of the second kind and $i = \sqrt{-1}$.

The formulation of the problem is the same as for the Laplace equation.

REFERENCES

1. BUTTERFIELD, R. and TOMLIN, G. R., "Integral techniques for solving zoned anisotropic problems", in *Variational Methods in Engineering, Vol.II*, Edited by C. A. Brebbia and H. Tottenham, Southampton University Press (1973)
2. CRUSE, T. A. and RIZZO, F. J., "A direct formulation and numerical solution of the general transient elastodynamic problem". *J. Math.. Anal. Applic.* Pt.1, **22**, 244 (1968)
3. CRUSE, T. A., "A direct formulation and numerical solution of the general transient elastodynamic problem", *J. Math. Anal. Applic.*, Pt.II, **22**, 341 (1968)
4. CRUSE, T. A., "Application of the boundary integral equation solution method in solid mechanics", in *Variational Methods in Engineering*, Edited by C. A. Brebbia and H. Tottenham, Southampton University Press (1973)
5. LACHAT, J. C., "A Further Development of the Boundary Integral Technique for Elastostatics", Ph.D. Thesis, University of Southampton (1973)

6. RIZZO, F. J. "An integral equation approach to boundary value problems of classical elastostatics", *Q. Appl. Math.*, **25**, No.83 (1967)
7. RIZZO, F. J. and SHIPPY, R., "A formulation and solution procedure for the general non-homogeneous elastic inclusion problem", *Int. J. Solids Struct*, **4**, 1161 (1968)
8. TOMLIN, G. R., "Numerical Analysis of Continuum Problems in Zoned Anisotropic Media", Ph.D. Thesis, University of Southampton (1973)
9. WATSON, J. O., "The Analysis of Thick Shells with Holes, by Integral Representation of Displacement", Ph.D. Thesis, University of Southampton (1972)
10. STROUD, A. H. and SECREST, D. *Gaussian Quadrature Formulae*, Prentice-Hall, New York (1966)
11. BREBBIA, C. A. and FERRANTE, A. *Computational Methods for the Solution of Engineering Problems*, Pentech Press, London (1978)
12. BREBBIA, C. A. and DOMINGUEZ, J. "Boundary elements versus finite elements", *Proc. International Conference on Applied Numerical Modelling*, Southampton University, July 1977, edited by C. A. Brebbia, Pentech Press, London (1978)
13. MARTIN, H. C., "Finite element analysis of fluid flows", *Proc. Conf. Matrix Methods in Struct. Mech.*, AFFDL TR 68-150, Patterson Air Force Base, Ohio, U.S.A. (1969)
14. CONNOR, J. J. and BREBBIA, C. A. *Finite Element Techniques for Fluid Flow*, 2nd ed., Newnes–Butterworth (1977)

4

Elasticity problems

4.1 INTRODUCTION

In what follows we will apply the boundary element method to linearly elastic problems. Within certain limits the behaviour of solids can be considered to be linear. In general, however, the materials are such that their properties are a function of time and the state of stress. Certain materials, for instance, creep with time and others present plastic modes of deformation at certain loads. In addition a body can crack, which produces a redistribution of stresses.

Even in the cases for which the material properties can be considered to be linear the deformations of the body may be such that its behaviour can no longer be assumed to be linear, i.e. the new geometry of the body will have to be taken into consideration.

All these problems can be attacked in one way or another using boundary elements but they are outside the scope of the book. We will restrict our discussion to linear-elastic systems.

The linear theory of elasticity for a solid body is based on the following assumptions:

(1) Linear material behaviour, i.e. linear stress–strain relations.
(2) The change in orientation of a body due to displacements is negligible. This assumption leads to linear strain displacement relations and also allows us to refer the equilibrium relations to the undeformed geometry.

In what follows we will use indicial notation as well as matrices for brevity and most relationships will apply equally well to two- or three-dimensional bodies.

4.2 LINEAR THEORY OF ELASTICITY

Let us first define the *state of stress* at a point (Figure 4.1). This is done by defining a stress tensor,

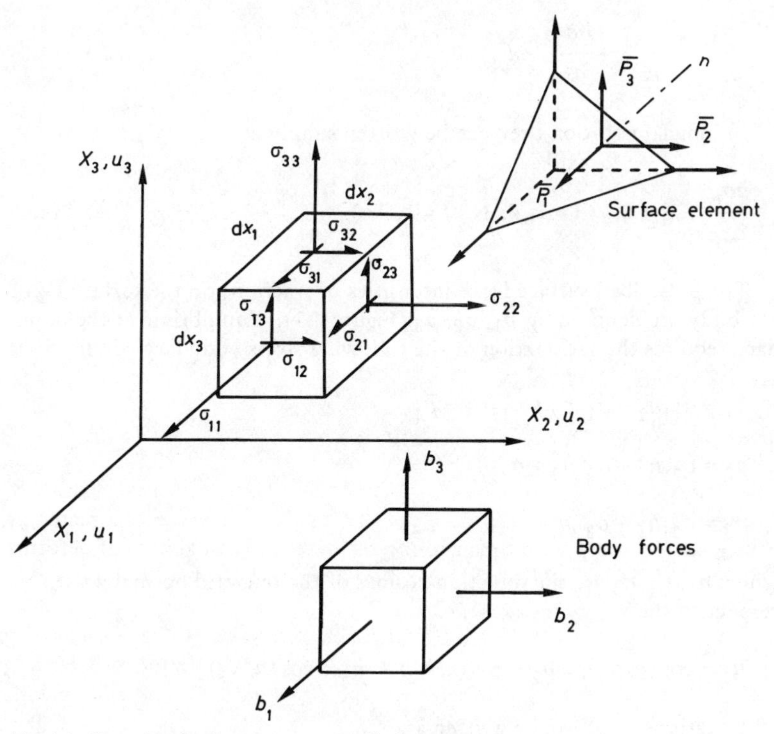

Fig.4.1. Notation for force, stress and displacement

$$\begin{bmatrix} \sigma_{11} & \sigma_{12} & \sigma_{13} \\ \sigma_{21} & \sigma_{22} & \sigma_{23} \\ \sigma_{31} & \sigma_{32} & \sigma_{33} \end{bmatrix} \tag{4.1}$$

where,

$$\sigma_{21} = \sigma_{12}, \quad \sigma_{31} = \sigma_{13}, \quad \sigma_{32} = \sigma_{23} \tag{4.2}$$

These stress components must satisfy the following equilibrium equations throughout the interior of the body,

$$\frac{\partial \sigma_{11}}{\partial x_1} + \frac{\partial \sigma_{12}}{\partial x_2} + \frac{\partial \sigma_{13}}{\partial x_3} + b_1 = 0$$

$$\frac{\partial \sigma_{21}}{\partial x_1} + \frac{\partial \sigma_{22}}{\partial x_2} + \frac{\partial \sigma_{23}}{\partial x_3} + b_2 = 0$$

$$\frac{\partial \sigma_{31}}{\partial x_1} + \frac{\partial \sigma_{32}}{\partial x_2} + \frac{\partial \sigma_{33}}{\partial x_3} + b_3 = 0 \tag{4.3}$$

In indicial notation they can be written simply as,

$$\frac{\partial \sigma_{ij}}{\partial x_j} + b_i = 0, \quad i = 1, 2, 3 \quad j = 1, 2, 3 \tag{4.4}$$

The prescribed surface force intensities or tractions on the surface Γ_2 of the body are denoted by $\bar{p}_1, \bar{p}_2, \bar{p}_3$ (Figure 4.1). Equilibrium at the boundary requires the satisfaction of the following stress boundary conditions,

$$p_1 = \sigma_{11}n_1 + \sigma_{12}n_2 + \sigma_{13}n_3 = \bar{p}_1$$

$$p_2 = \sigma_{21}n_1 + \sigma_{22}n_2 + \sigma_{23}n_3 = \bar{p}_2$$

$$p_3 = \sigma_{31}n_1 + \sigma_{32}n_2 + \sigma_{33}n_3 = \bar{p}_3 \tag{4.5}$$

where n_1, n_2, n_3 are the direction cosines of the outward normal n with respect to the x_1, x_2, x_3 axes, i.e.

$$n_1 = \cos(n, x_1), \quad n_2 = \cos(n, x_2), \quad n_3 = \cos(n, x_3)$$

Equations (4.5) can be written as,

$$p_i = \sigma_{ij}n_j = \bar{p}_i, \quad i = 1, 2, 3 \quad j = 1, 2, 3 \tag{4.6}$$

The state of strain at a point is defined by the following strain tensor,

$$\begin{bmatrix} \epsilon_{11} & \epsilon_{12} & \epsilon_{13} \\ \epsilon_{21} & \epsilon_{22} & \epsilon_{23} \\ \epsilon_{31} & \epsilon_{32} & \epsilon_{33} \end{bmatrix} \tag{4.7}$$

where $\epsilon_{21} = \epsilon_{12}$, $\epsilon_{31} = \epsilon_{13}$, $\epsilon_{32} = \epsilon_{23}$. The strain-displacement relations for the linear theory are,

$$\epsilon_{11} = \frac{\partial u_1}{\partial x_1}, \quad \epsilon_{22} = \frac{\partial u_2}{\partial x_2}, \quad \epsilon_{33} = \frac{\partial u_3}{\partial x_3} \tag{4.8}$$

$$\epsilon_{12} = \frac{1}{2}\left(\frac{\partial u_1}{\partial x_2} + \frac{\partial u_2}{\partial x_1}\right), \quad \epsilon_{13} = \frac{1}{2}\left(\frac{\partial u_1}{\partial x_3} + \frac{\partial u_3}{\partial x_1}\right), \quad \epsilon_{23} = \frac{1}{2}\left(\frac{\partial u_2}{\partial x_3} + \frac{\partial u_3}{\partial x_2}\right)$$

This expression can also be written in indicial form as,

$$\epsilon_{ij} = \frac{1}{2}\left(\frac{\partial u_i}{\partial x_j} + \frac{\partial u_j}{\partial x_i}\right), \quad i = 1, 2, 3 \quad j = 1, 2, 3 \tag{4.9}$$

Let Γ_1 denote the portion of the boundary on which displacements are prescribed. The displacement constraints there are,

$$u_1 = \overline{u}_1, \quad u_2 = \overline{u}_2, \quad u_3 = \overline{u}_3, \quad \text{on } \Gamma_1$$

or

$$u_j = \overline{u}_j \qquad j = 1, 2, 3 \tag{4.10}$$

The $\overline{u}_j$ are the prescribed values. Note that the total surface Γ of the body is equal to $\Gamma_1 + \Gamma_2$.

The states of stress and strain for a body are related throughout the stress–strain relations, which depend on the material behaviour. For a linearly elastic material we have in matrix form

$$\begin{Bmatrix} \epsilon_{11} \\ \epsilon_{22} \\ \epsilon_{33} \\ 2\epsilon_{12} \\ 2\epsilon_{13} \\ 2\epsilon_{23} \end{Bmatrix} = \begin{bmatrix} c_{11} & c_{12} & c_{13} & c_{14} & c_{15} & c_{16} \\ & c_{22} & c_{23} & c_{24} & c_{25} & c_{26} \\ & & c_{33} & c_{34} & c_{35} & c_{36} \\ & \text{sym.} & & c_{44} & c_{45} & c_{46} \\ & & & & c_{55} & c_{56} \\ & & & & & c_{66} \end{bmatrix} \begin{Bmatrix} \sigma_{11} \\ \sigma_{22} \\ \sigma_{33} \\ \sigma_{12} \\ \sigma_{13} \\ \sigma_{23} \end{Bmatrix} \tag{4.11}$$

or

$$\epsilon = \mathbf{C}\sigma$$

where c_{ij} are called elastic compliances.

We express the inverted form of Equation (4.11) as,

$$\begin{Bmatrix} \sigma_{11} \\ \sigma_{22} \\ \sigma_{33} \\ \sigma_{12} \\ \sigma_{13} \\ \sigma_{23} \end{Bmatrix} = \begin{bmatrix} d_{11} & d_{12} & d_{13} & d_{14} & d_{15} & d_{16} \\ & d_{22} & d_{23} & d_{24} & d_{25} & d_{26} \\ & & d_{33} & d_{34} & d_{35} & d_{36} \\ & & & d_{44} & d_{45} & d_{46} \\ & \text{sym.} & & & d_{55} & d_{56} \\ & & & & & d_{66} \end{bmatrix} \begin{Bmatrix} \epsilon_{11} \\ \epsilon_{22} \\ \epsilon_{33} \\ 2\epsilon_{12} \\ 2\epsilon_{13} \\ 2\epsilon_{23} \end{Bmatrix} \tag{4.12}$$

or

$$\sigma = \mathbf{D}\epsilon$$

d_{ij} are called rigidity coefficients.

There are 21 material constants for an elastic (Green-type) material. The number of independent constants is reduced when the material structure has one or more planes of symmetry. If the material has three orthogonal planes of symmetry it is said to be orthotropic, and Equation (4.11) reduces to,

$$\begin{Bmatrix} \epsilon_{11} \\ \epsilon_{22} \\ \epsilon_{33} \\ 2\epsilon_{12} \\ 2\epsilon_{13} \\ 2\epsilon_{23} \end{Bmatrix} = \begin{bmatrix} c_{11} & c_{12} & c_{13} & & & \\ & c_{22} & c_{23} & & 0 & \\ & & c_{33} & & & \\ & & & c_{44} & 0 & 0 \\ \text{sym.} & & & & c_{55} & 0 \\ & & & & & c_{66} \end{bmatrix} \begin{Bmatrix} \sigma_{11} \\ \sigma_{22} \\ \sigma_{33} \\ \sigma_{12} \\ \sigma_{13} \\ \sigma_{23} \end{Bmatrix} \tag{4.13}$$

where $x_1 - x_2$, $x_1 - x_3$ and $x_2 - x_3$ are material symmetry planes. The inverted relations are similar in form to (4.13).

An isotropic material has only two independent constants. By definition, the form of the stress–strain relations is invariant, i.e. independent of the choice of reference frame. The form of Equation (4.12) is,

$$\begin{Bmatrix} \sigma_{11} \\ \sigma_{22} \\ \sigma_{33} \\ \sigma_{12} \\ \sigma_{13} \\ \sigma_{23} \end{Bmatrix} = \frac{E}{2(1+\nu)} \begin{bmatrix} \frac{2(1-\nu)}{(1-2\nu)} & \frac{2\nu}{1-2\nu} & \frac{2\nu}{1-2\nu} & & 0 & \\ & \frac{2(1-\nu)}{1-2\nu} & \frac{2\nu}{1-2\nu} & & & \\ & & \frac{2(1-\nu)}{1-2\nu} & & & \\ & & & 1 & 0 & 0 \\ & \text{sym.} & & & 1 & 0 \\ & & & & & 1 \end{bmatrix} \begin{Bmatrix} \epsilon_{11} \\ \epsilon_{22} \\ \epsilon_{33} \\ 2\epsilon_{12} \\ 2\epsilon_{13} \\ 2\epsilon_{23} \end{Bmatrix} \tag{4.14}$$

where E and ν are Young's modulus and Poisson's ratio respectively. The inverted relations are generally written as,

$$\begin{Bmatrix} \epsilon_{11} \\ \epsilon_{22} \\ \epsilon_{33} \\ 2\epsilon_{12} \\ 2\epsilon_{13} \\ 2\epsilon_{23} \end{Bmatrix} = \frac{1}{E} \begin{bmatrix} 1 & -\nu & -\nu & & & \\ & 1 & -\nu & & 0 & \\ & & 1 & & & \\ & & & 2(1+\nu) & 0 & 0 \\ \text{sym.} & & & & 2(1+\nu) & 0 \\ & & & & & 2(1+\nu) \end{bmatrix} \begin{Bmatrix} \sigma_{11} \\ \sigma_{22} \\ \sigma_{33} \\ \sigma_{12} \\ \sigma_{13} \\ \sigma_{23} \end{Bmatrix} \tag{4.15}$$

The relations (4.14) and (4.15) can also be expressed in terms of Lame's constants (λ, μ), which are related to E and ν in the following way.

$$\lambda = E\nu/[(1+\nu)\,(1-2\nu)]$$

$$\mu = E/[2(1+\nu)] = G \tag{4.16}$$

Initial strains or stresses

In many problems one has initial strains or stresses, due to temperature or other causes. In these cases the strain–stress relationships can be written as,

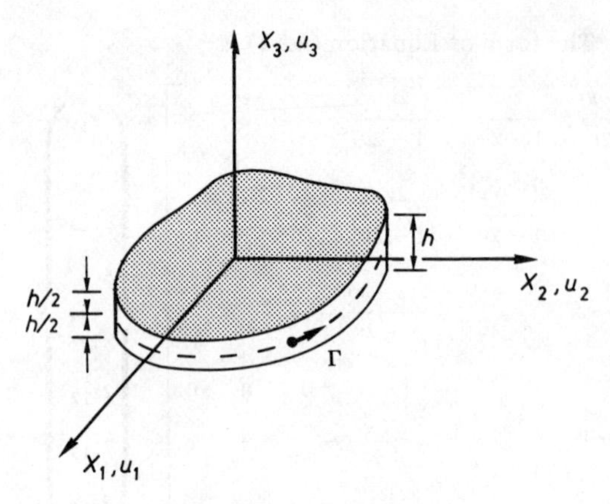

Fig.4.2. Prismatic solid

$$(\boldsymbol{\epsilon} - \boldsymbol{\epsilon}^0) = \mathbf{C}\boldsymbol{\sigma} \tag{4.17}$$

where $\boldsymbol{\epsilon}^0$ is the initial strain vector, $\boldsymbol{\epsilon}$ is the total strain vector and the difference $(\boldsymbol{\epsilon} - \boldsymbol{\epsilon}^0)$ represents the elastic part.

One can also work in terms of initial stresses $\boldsymbol{\sigma}^0$ and inverting Equation (4.17) we have that,

$$\begin{aligned} \boldsymbol{\sigma} &= \mathbf{D}(\boldsymbol{\epsilon} - \boldsymbol{\epsilon}^0) = \mathbf{D}\boldsymbol{\epsilon} - \mathbf{D}\boldsymbol{\epsilon}^0 \\ &= \mathbf{D}\boldsymbol{\epsilon} + \boldsymbol{\sigma}^0 \end{aligned} \tag{4.18}$$

where $\boldsymbol{\sigma}^0 = -\mathbf{D}\boldsymbol{\epsilon}^0$ is the initial stress vector.

It will be shown in Section 4.5 that the effect of initial stress or strain fields can be analysed in the same way as the body forces.

Plate stretching and plane strain

Consider the prismatic homogeneous solid shown in Figure 4.2. The end surfaces are defined by $x_3 = \pm h/2$ and the cylindrical surface by $x_1 = x_1(\Gamma)$, $x_2 = x_2(\Gamma)$, where Γ is the arc length along the boundary curve. If,

(1) the body is thin (h is small in comparison with the representative dimensions in $x_1 x_2$ directions),

(2) there are no surface forces acting on the end faces ($p_1 = p_2 = p_3 = 0$ at $x_3 = \pm h/2$),

(3) the body forces are planar and independent of x_3 ($b_3 = 0$ and b_1, b_2 are functions only of $x_1\, x_2$), and

(4) the forces acting on the cylindrical boundary are planar and independent of x_3 ($p_3 = 0$ and p_1, p_2 are functions only of $x_1\, x_2$)

then it is reasonable to assume that σ_{33}, σ_{31}, σ_{23} are small in comparison with σ_{11} σ_{22} σ_{12} and the variation of all these quantities with respect to x_3 is small. The plane stress or plate stretching problem is based on the following assumptions,

$$\sigma_{33} = \sigma_{31} = \sigma_{32} = 0$$

throughout the volume and

$$\sigma_{11}, \sigma_{22}, \sigma_{12} \text{ are independent of } x_3. \tag{4.19}$$

(Rigorously requiring σ_{11} σ_{22} σ_{12} to be independent of x_3 violates certain compatibility equations but it is quite reasonable for thin plates.)

An alternative statement of Equation (4.19) is

$$u_1 = u_1(x_1, x_2); \quad u_2 = u_2(x_1, x_2)$$

$$\sigma_{33} = \sigma_{31} = \sigma_{32} = 0 \tag{4.20}$$

We take Equation (4.20) as the basic assumptions for plate stretching. Equations (4.19) apply for a homogeneous plate. In order to treat a laminated plate, u_1 and u_2 are assumed to be constant throughout the thickness and the out-of-plane stresses are neglected. The in-plane stresses vary according to the material properties.

The transverse displacement u_3 is not independent and can be obtained from the stress–strain relation for ϵ_{33}:

$$\epsilon_{33} = \frac{\partial u_3}{\partial x_3} = \text{function of } \sigma_{11}, \sigma_{22}, \sigma_{12} \tag{4.21}$$

The other possibility, 'plane strain', is to start by assuming that the u_3 displacement is zero. This can be done if

(1) The thickness is large in comparison with the representative dimensions in $x_1 x_2$ directions.
(2) The body forces and surface forces acting on the cylindrical surface are planar (no x_3 component) and independent of x_3.

It is then reasonable to assume that the in-plane displacements u_1 and u_2 are independent of x_3, and the out-of-plane displacement u_3 is zero. The 'plane strain' problem is based on the following assumptions,

$$u_1 = u_1(x_1, x_2), \quad u_2 = u_2(x_1, x_2), \quad u_3 = 0 \tag{4.22}$$

Equation (4.22) corresponds to specifying

$$\epsilon_{33} = \epsilon_{31} = \epsilon_{32} = 0$$

$$\epsilon_{11}, \epsilon_{22}, \epsilon_{12} \text{ are independent of } x_3 \tag{4.23}$$

Note that $\sigma_{33} \neq 0$ for plane strain. One can determine σ_{33} from the stress–strain relationship for σ_{33}.

Both plate stretching and plane strain are two dimensional problems in the sense that the displacement and force parameters vary only with respect to x_1 and x_2.

Let us now formulate the two dimensional stress–strain relations for plate stretching and plane strain. We start with the three-dimensional relations for the linear elastic material, i.e.

$$\boldsymbol{\epsilon} - \boldsymbol{\epsilon}^0 = \mathbf{C}\boldsymbol{\sigma} \tag{4.24}$$

where $\boldsymbol{\epsilon}^0$ are the initial strain components. Relationship (4.24) involves 21 different constant c's. For the problem to be two-dimensional, the x_1–x_2 plane must be a plane of symmetry for the material structure, i.e. the stress–strain relations must have the same form when the x_3 direction is reversed. One can easily show that the appropriate relations are,

$$\begin{Bmatrix} \epsilon_{11}-\epsilon_{11}^0 \\ \epsilon_{22}-\epsilon_{22}^0 \\ \epsilon_{33}-\epsilon_{33}^0 \\ 2(\epsilon_{12}-\epsilon_{12}^0) \\ 2\epsilon_{13} \\ 2\epsilon_{23} \end{Bmatrix} = \begin{bmatrix} c_{11} & c_{12} & c_{13} & c_{14} & 0 & 0 \\ c_{12} & c_{22} & c_{23} & c_{24} & 0 & 0 \\ c_{13} & c_{23} & c_{33} & c_{34} & 0 & 0 \\ c_{14} & c_{24} & c_{34} & c_{44} & 0 & 0 \\ 0 & 0 & 0 & 0 & c_{55} & c_{56} \\ 0 & 0 & 0 & 0 & c_{56} & c_{66} \end{bmatrix} \begin{Bmatrix} \sigma_{11} \\ \sigma_{22} \\ \sigma_{33} \\ \sigma_{12} \\ \sigma_{13} \\ \sigma_{23} \end{Bmatrix} \tag{4.25}$$

If either x_1 or x_2 is an axis of symmetry, the material is said to be orhtotropic. For this case,

$$c_{14} = c_{24} = c_{34} = c_{56} = 0$$

$$\epsilon_{12}^0 = 0 \tag{4.26}$$

If the initial strains are due to temperature we have

$$\epsilon_{11}^0 = \alpha_1 \Delta T, \quad \epsilon_{22}^0 = \alpha_2 \Delta T, \quad \epsilon_{33}^0 = \alpha_3 \Delta T \tag{4.27}$$

where ΔT is the temperature increment and α_1, α_2, α_3 are the thermal strain coefficients. Note that Equations (4.27) apply only when x_1 and x_2 are material symmetry directions, i.e. orthotropic directions.

Additional simplifications can be obtained by introducing further restrictions, such as equal properties in the two orthotropic directions or any two directions in a plane (plane isotropy). For more detail see reference [1].

The relations for *plane stress* are obtained by setting $\sigma_{33} = 0$ in (4.25) and considering only the equations for ϵ_{11}, ϵ_{22}, ϵ_{12}, i.e.

$$\boldsymbol{\epsilon} = \boldsymbol{\epsilon}^0 + \mathbf{C}\boldsymbol{\sigma} \tag{4.28}$$

or

$$\begin{Bmatrix} \epsilon_{11} \\ \epsilon_{22} \\ 2\epsilon_{12} \end{Bmatrix} = \begin{Bmatrix} \epsilon_{11}^0 \\ \epsilon_{22}^0 \\ 2\epsilon_{12}^0 \end{Bmatrix} + \begin{bmatrix} c_{11} & c_{12} & c_{14} \\ c_{12} & c_{22} & c_{24} \\ c_{14} & c_{24} & c_{44} \end{bmatrix} \begin{Bmatrix} \sigma_{11} \\ \sigma_{22} \\ \sigma_{12} \end{Bmatrix} \tag{4.29}$$

Solving for $\boldsymbol{\sigma}$ leads to

$$\boldsymbol{\sigma} = \mathbf{C}^{-1}(\boldsymbol{\epsilon} - \boldsymbol{\epsilon}^0) = \boldsymbol{\sigma}^0 + \mathbf{D}\boldsymbol{\epsilon} \tag{4.30}$$

For *plane strain* we must first eliminate σ_{33} using the conditions,

$$\epsilon_{33} = \epsilon_{33}^0 + c_{13}\sigma_{11} + c_{23}\sigma_{22} + c_{33}\sigma_{33} + c_{34}\sigma_{12} = 0 \tag{4.31}$$

This gives,

$$\sigma_{33} = -\frac{1}{c_{33}} [\epsilon_{33}^0 + c_{13}\sigma_{11} + c_{23}\sigma_{22} + c_{34}\sigma_{12}] \tag{4.32}$$

Substituting for σ_{33} in Equation (4.25) we obtain,

$$\boldsymbol{\epsilon} = \boldsymbol{\epsilon}_0' + \mathbf{C}'\boldsymbol{\sigma} \tag{4.33}$$

or

$$\begin{Bmatrix} \epsilon_{11} \\ \epsilon_{22} \\ 2\epsilon_{12} \end{Bmatrix} = \begin{Bmatrix} \epsilon_{11}^0 \\ \epsilon_{22}^0 \\ 2\epsilon_{12}^0 \end{Bmatrix} - \epsilon_{33}^0 \begin{Bmatrix} c_{13}/c_{33} \\ c_{23}/c_{33} \\ c_{34}/c_{33} \end{Bmatrix}$$

$$+ \begin{bmatrix} c_{11} - (c_{13}^2/c_{33}) & c_{12} - (c_{13}c_{23}/c_{33}) & c_{14} - (c_{13}c_{34}/c_{33}) \\ & c_{22} - (c_{23}^2/c_{33}) & c_{24} - (c_{23}c_{34}/c_{33}) \\ \text{sym.} & & c_{44} - (c_{34}^2/c_{33}) \end{bmatrix} \begin{Bmatrix} \sigma_{11} \\ \sigma_{22} \\ \sigma_{12} \end{Bmatrix} \tag{4.34}$$

Finally we invert Equation (4.34)

$$\boldsymbol{\sigma} = (\mathbf{C}')^{-1}(\boldsymbol{\epsilon} - \boldsymbol{\epsilon}_0') = \boldsymbol{\sigma}_0' + \mathbf{D}'\boldsymbol{\epsilon} \tag{4.35}$$

We could have obtained Equation (4.35) by first inverting (4.25) and then setting $\epsilon_{33} = 0$.

In general, $\mathbf{D}$ is determined by inverting $\mathbf{C}$ ($\mathbf{D}'$ and $\mathbf{C}'$ for plane strain). When the material is *orthotropic* and the material symmetry directions coincide with the x_1, x_2 directions, one can readily obtain the explicit form for $\mathbf{D}$. Rather than use the c_{ij} notation as in Equation (4.25) we express the stress–strain relations in terms of the physical constants, i.e. the extensional and shear moduli and the Poisson coefficients, and write Equation (4.25) as,

$$\begin{aligned} \epsilon_{11} &= \frac{1}{E_{11}}\sigma_{11} - \frac{\nu_{21}}{E_{22}}\sigma_{22} - \frac{\nu_{31}}{E_{33}}\sigma_{33} + \epsilon_{11}^0 \\ \epsilon_{22} &= -\frac{\nu_{12}}{E_{11}}\sigma_{11} + \frac{1}{E_{22}}\sigma_{22} - \frac{\nu_{32}}{E_{33}}\sigma_{33} + \epsilon_{22}^0 \\ \epsilon_{33} &= -\frac{\nu_{31}}{E_{11}}\sigma_{11} - \frac{\nu_{32}}{E_{22}}\sigma_{22} + \frac{1}{E_{33}}\sigma_{33} + \epsilon_{33}^0 \\ 2\epsilon_{12} &= \frac{1}{G_{12}}\sigma_{12} \end{aligned} \tag{4.36}$$

Specializing Equation (4.36) for plane stress ($\sigma_{33} = 0$) and inverting leads to,

$$\boldsymbol{\sigma} = \mathbf{D}(\boldsymbol{\epsilon} - \boldsymbol{\epsilon}^0) \tag{4.37}$$

That is

$$\begin{Bmatrix} \sigma_{11} \\ \sigma_{22} \\ \sigma_{12} \end{Bmatrix} = \begin{bmatrix} \dfrac{E_{11}}{1 - n\nu_{21}^2} & \dfrac{\nu_{21}E_{11}}{1 - n\nu_{21}^2} & 0 \\ & \dfrac{E_{11}}{n(1 - n\nu_{21}^2)} & 0 \\ \text{sym.} & & G_{12} \end{bmatrix} \begin{Bmatrix} \epsilon_{11} - \epsilon_{11}^0 \\ \epsilon_{22} - \epsilon_{22}^0 \\ 2\epsilon_{12} \end{Bmatrix} \tag{4.38}$$

where

$$n = \frac{E_{11}}{E_{22}} .$$

The relations for an isotropic material follow by setting

$$E_{11} = E_{22} = E, \quad \nu_{21} = \nu, \quad G_{12} = \frac{E}{2(1 + \nu)} \tag{4.39}$$

(plus $\alpha = \alpha_{11} = \alpha_{22}$ for the thermal strain coefficient).
For plane strain we have that,

$$\sigma_{33} = \nu_{31}\sigma_{11} + \nu_{32}\sigma_{22} - E_{33}\epsilon_{33}^0 \tag{4.40}$$

The remaining relations in Equation (4.36) can be written as,

$$\epsilon_{11} = \frac{1}{E_{11}'}\sigma_{11} - \frac{\nu_{21}'}{E_{11}'}\sigma_{22} + (\epsilon_{11}^0 + \nu_{31}\epsilon_{33}^0)$$

$$\epsilon_{22} = -\frac{\nu_{21}'}{E_{22}'}\sigma_{22} + \frac{1}{E_{22}'}\sigma_{22} + (\epsilon_{22}^0 + \nu_{32}\epsilon_{33}^0)$$

$$2\epsilon_{12} = \frac{1}{G_{12}}\sigma_{12} \tag{4.41}$$

where,

$$E_{11}' = \frac{E_{11}}{1 - (E_{11}/E_{22})\nu_{31}^2}$$

$$E'_{22} = \frac{E_{22}}{1 - (E_{22}/E_{11})\nu_{32}^2}$$

$$\nu'_{21} = \frac{\nu_{21} + \nu_{31}\nu_{32}(E_{22}/E_{33})}{1 - \nu_{32}^2(E_{22}/E_{33})} \tag{4.42}$$

Note that for temperature problems the α' coefficients can be defined as

$$\alpha'_1 = \alpha_1 + \nu_{31}\alpha_3$$

$$\alpha'_2 = \alpha_2 + \nu_{32}\alpha_3$$

Since the expressions have the same forms as for plane stress, one just has to replace E_{11}, E_{22}, ν_{21}, with $E'_{11}, E'_{22}, \nu'_{21}$ in Equation (4.38) to obtain $\mathbf{D}'$ and $\boldsymbol{\sigma}'^0$. Note that now,

$$n' = \frac{E'_{11}}{E'_{22}} = \frac{E_{11}}{E_{22}} \left[\frac{1 - (E_{22}/E_{33})\nu_{32}^2}{1 - (E_{11}/E_{33})\nu_{31}^2} \right] \tag{4.43}$$

The equivalent coefficients for the isotropic case are,

$$E'_{11} = \boldsymbol{E}'_{33} = E' = E/(1 - \nu^2)$$

$$\nu'_{12} = \nu' = \nu/(1 - \nu)$$

$$G_{12} = E/(2(1 + \nu)) \tag{4.44}$$

For the thermal strain coefficients we have,

$$\alpha'_1 = \alpha'_2 = \alpha(1 + \nu) \tag{4.45}$$

Since plane stress and plane strain differ only in the elements of the rigidity and initial strain matrices, we have to develop only one formulation. In what follows, we will not distinguish between the two problems.

4.3 BASIC RELATIONSHIPS

The principle of virtual displacements for linear elastic problems can be written as,

$$\int_{\Omega} (\sigma_{jk,j} + b_k)u_k^* d\Omega = \int_{\Gamma_2} (p_k - \bar{p}_k)u_k^* d\Gamma \tag{4.46}$$

where u_k^* are the virtual displacements identically satisfying the homogeneous boundary conditions $\bar{u}_k^* \equiv 0$ on Γ_1. If we now interpret u_k^* as weighting functions which do not satisfy these conditions on Γ_1 the expression can be written as,

$$\int_\Omega (\sigma_{jk,j} + b_k)u_k^* d\Omega = \int_{\Gamma_2} (p_k - \bar{p}_k)u_k^* d\Gamma + \int_{\Gamma_1} (\bar{u}_k - u_k)p_k^* d\Gamma \tag{4.47}$$

where $p_k^* = n_j\sigma_{jk}^*$ are the surface forces or tractions corresponding to the u_k^* system. We will assume that the strain-displacement relationships are linear, i.e.

$$\epsilon_{ij} = \frac{1}{2}\left(\frac{\partial u_i}{\partial x_j} + \frac{\partial u_j}{\partial x_i}\right) \tag{4.48}$$

and

$$\epsilon_{ij}^* = \frac{1}{2}\left(\frac{\partial u_i^*}{\partial x_j} + \frac{\partial u_j^*}{\partial x_i}\right) \tag{4.49}$$

and that the material properties are also linear.

Hence we can now integrate Equation (4.47) by parts giving

$$\int_\Omega b_k u_k^* d\Omega - \int_\Omega \sigma_{jk}\epsilon_{jk}^* d\Omega = -\int_{\Gamma_2} \bar{p}_k u_k^* d\Gamma - \int_{\Gamma_1} p_k u_k^* d\Gamma$$
$$+ \int_{\Gamma_1} (\bar{u}_k - u_k)p_k^* d\Gamma \tag{4.50}$$

Integrating by parts once more we obtain,

$$\int_\Omega b_k u_k^* d\Omega + \int_\Omega \sigma_{jk,j}^* u_k d\Omega = -\int_{\Gamma_2} \bar{p}_k u_k^* d\Gamma - \int_{\Gamma_1} p_k u_k^* d\Gamma$$
$$+ \int_{\Gamma_1} \bar{u}_k p_k^* d\Gamma + \int_{\Gamma_2} u_k p_k^* d\Gamma \tag{4.51}$$

We now look for fundamental solutions satisfying the equilibrium equations, usually of the type,

$$\sigma_{jk,j}^* + \Delta_l^i = 0 \tag{4.52}$$

where Δ_l^i is the Dirac delta function and represents a unit load at i in the l direction. This type of solution will produce for each direction 'l' the following equation,

$$u_l^i + \int_{\Gamma_1} \bar{u}_k p_k^* \mathrm{d}\Gamma + \int_{\Gamma_2} u_k p_k^* \mathrm{d}\Gamma = \int_{\Omega} b_k u_k^* \mathrm{d}\Omega + \int_{\Gamma_1} p_k u_k^* \mathrm{d}\Gamma$$

$$+ \int_{\Gamma_2} \bar{p}_k u_k^* \mathrm{d}\Gamma \tag{4.53}$$

u_l^i represents the displacement at i in the 'l' direction.

In general we can write for the point 'i'

$$u_l^i + \int_{\Gamma} u_k p_k^* \mathrm{d}\Gamma = \int_{\Gamma} p_k u_k^* \mathrm{d}\Gamma + \int_{\Omega} b_k u_k^* \mathrm{d}\Omega \tag{4.54}$$

where $\Gamma = \Gamma_1 + \Gamma_2$.

Note that u_k^* and p_k^* are the fundamental solutions, i.e. the displacements and tractions due to a unit concentrated load at the point 'i' in the 'l' direction. If we consider unit forces acting in the three directions, Equation (4.54) can be written as,

$$u_l^i + \int_{\Gamma} u_k p_{lk}^* \mathrm{d}\Gamma = \int_{\Gamma} p_k u_{lk}^* \mathrm{d}\Gamma + \int_{\Omega} b_k u_{lk}^* \mathrm{d}\Omega \tag{4.55}$$

where p_{lk}^* and u_{lk}^* represent the tractions and displacements in the k direction due to unit forces acting in the l direction. Equation (4.55) is valid for the particular point 'i' where these forces are applied.

Fundamental solutions

The fundamental solution for a three-dimensional isotropic body is

$$u_{lk}^* = \frac{1}{16\pi G(1-\nu)} \frac{1}{r}\left[(3-4\nu)\Delta_{lk} + \frac{\partial r}{\partial x_l}\frac{\partial r}{\partial x_k}\right] \tag{4.56}$$

$$p_{lk}^* = -\frac{1}{8\pi(1-\nu)r^2}\left[\frac{\partial r}{\partial n}\left\{(1-2\nu)\Delta_{lk} + 3\frac{\partial r}{\partial x_l}\frac{\partial r}{\partial x_k}\right\}\right.$$

$$\left. - (1-2\nu)\left\{\frac{\partial r}{\partial x_l} n_k - \frac{\partial r}{\partial x_k} n_l\right\}\right]$$

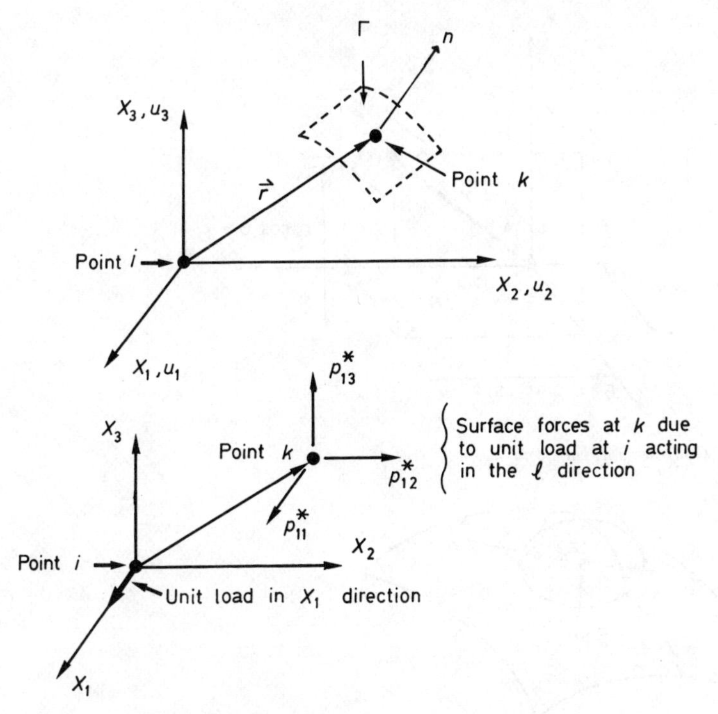

Fig.4.3. Three-dimensional case

where n is the outward unit normal to the surface of the body, Δ_{lk} is the Kronecker delta, r is the distance from the point of application of the load to the point under consideration and n_j are the direction cosines (see Figure 4.3).

For two dimensional isotropic plane strain case the solution is,

$$u^*_{lk} = \frac{1}{8\pi G(1-\nu)} \left[(3-4\nu) \ln \left(\frac{1}{r} \right) \Delta_{lk} + \frac{\partial r}{\partial x_l} \frac{\partial r}{\partial x_k} \right]$$

$$p^*_{lk} = -\frac{1}{4\pi(1-\nu)r} \left[\frac{\partial r}{\partial n} \left\{ (1-2\nu)\Delta_{kl} + 2 \frac{\partial r}{\partial x_k} \frac{\partial r}{\partial x_l} \right\} \right.$$

$$\left. - (1-2\nu) \left(\frac{\partial r}{\partial x_l} n_k - \frac{\partial r}{\partial x_k} n_l \right) \right] \tag{4.57}$$

where $(\partial r/\partial x_l) = (r_l/r)$ (see Figure 4.4).

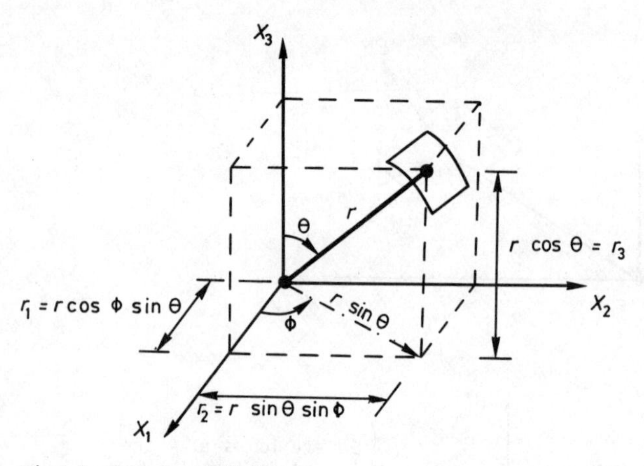

Fig.4.4. Geometry definitions

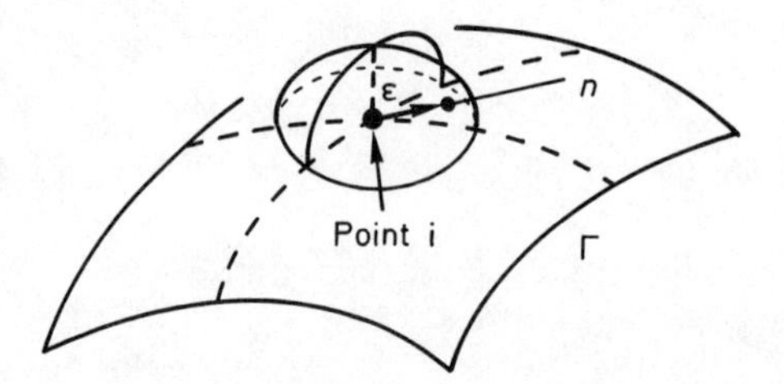

Fig.4.5. Boundary surface Γ_ϵ assumed hemispherical for integration purposes

Boundary point

As for the potential problem, Equation (4.55) will now be specialized for the boundary. Consider that the boundary is smooth and of the type Γ_2 at the point 'i' (the same will apply for Γ_1 boundary). The first integral on Γ_2 in Equation (4.55) can now be written into two parts,

$$\int_{\Gamma_2} u_k p_{lk}^* d\Gamma = \int_{\Gamma_{(2-\epsilon)}} u_k p_{lk}^* d\Gamma + \int_{\Gamma_\epsilon} u_k p_{lk}^* d\Gamma \tag{4.58}$$

Let us now consider the Γ_ϵ integral for the case $\epsilon \to 0$, (Figure 4.5) and call it I for simplicity,

$$I = \lim_{\epsilon\to 0} \left\{ \int_{S_\epsilon} u_k p_{lk}^* d\Gamma \right\}$$

$$= \lim_{\epsilon \to 0} \left\{ -\int_{\Gamma_\epsilon} u_k \left[\frac{\partial r}{\partial n} \left\{ (1-2\nu)\Delta_{lk} + 3\frac{\partial r}{\partial x_l}\frac{\partial r}{\partial x_k} \right\} \right.\right.$$

$$\left.\left. - (1-2\nu)\left\{ \frac{\partial r}{\partial x_l} n_k - \frac{\partial r}{\partial x_k} n_l \right\} \right] \frac{1}{8\pi(1-\nu)r^2} \right\} \mathrm{d}\Gamma \qquad (4.59)$$

Note that $\epsilon \equiv r$. Consider Figure 4.4 where for simplicity a spherical system of coordinates is used. For this particular case the second term in Equation (4.59) will disappear as,

$$\frac{\partial r}{\partial x_l} n_k - \frac{\partial r}{\partial x_k} n_l = \frac{\partial r}{\partial x_l}\frac{\partial x_k}{\partial r} - \frac{\partial r}{\partial x_k}\frac{\partial x_l}{\partial r} \equiv 0 \qquad (4.60)$$

Hence we only need to consider the first term in the integral, that is,

$$I = \lim_{\epsilon \to 0} \left\{ -\int_{\Gamma_\epsilon} u_k \frac{\partial r}{\partial n} \left\{ (1-2\nu)\Delta_{lk} + 3\frac{\partial r}{\partial x_l}\frac{\partial r}{\partial x_k} \right\} \frac{\mathrm{d}\Gamma}{8\pi(1-\nu)r^2} \right\} \qquad (4.61)$$

Note that $\frac{\partial r}{\partial n} = 1$

This can be expanded taking into account the geometric relationships shown in Figure 4.4, for instance when $l = 1$

$$I = \lim_{\epsilon \to 0} \left\{ -\int_{\Gamma_\epsilon} \{ u_1^i(1-2\nu) + 3u_1^i e_1 e_1 + 3u_2^i e_1 e_2 \right.$$

$$\left. + 3u_3^i e_1 e_3 \} \cdot \frac{\sin\theta \, \mathrm{d}\theta \, \mathrm{d}\phi}{8\pi(1-\nu)} \right. \qquad (4.62)$$

The e_i are unit vectors in the x_i direction (see Figure 4.4) such that

$$e_i = n_i = \frac{\partial r}{\partial x_i} = \frac{r_i}{r}$$

Note that the integral is now independent of 'r' and can be expressed in terms of ϕ and θ only,

$$I = -\int_0^{2\pi}\int_0^{\pi/2} \{ u_1^i(1-2\nu) + 3u_1^i \sin^2\theta \cos^2\phi$$

$$+ 3u_2^i \sin^2\theta \cos\phi \sin\phi$$

$$+ 3u_3^i \sin\theta \cos\theta \cos\phi \} \cdot \frac{\sin\theta \, \mathrm{d}\theta \, \mathrm{d}\phi}{8\pi(1-\nu)} \qquad (4.63)$$

After integration we find that some of the integrals are zero and the final result is,

$$I = -\frac{1}{8\pi(1-\nu)}[(1-2\nu)2\pi + 2\pi]\,u_1^i$$

$$= -\frac{4(1-\nu)}{8(1-\nu)}u_1^i = -\frac{1}{2}u_1^i \tag{4.64}$$

The same can be shown to apply for $l = 2$ and $l = 3$. This result can then be written as,

$$\lim_{\epsilon\to 0}\left\{\int_{\Gamma_\epsilon} u_k p_{lk}^* \mathrm{d}\Gamma\right\} = -\frac{u_l^i}{2} \tag{4.65}$$

The integral

$$\int_{\Gamma_2} \bar{p}_k u_{lk}^* \mathrm{d}\Gamma$$

can also be written

$$\int_{\Gamma_{(2-\epsilon)}} \bar{p}_k u_{lk}^* \mathrm{d}\Gamma + \int_{\Gamma_\epsilon} \bar{p}_k u_{lk}^* \mathrm{d}\Gamma$$

but it can be easily be shown that

$$\lim_{\epsilon\to 0}\int_{\Gamma_\epsilon} \bar{p}_k u_{lk}^* \mathrm{d}\Gamma = 0$$

and therefore this integral does not introduce any new term.

A similar result would be obtained for the Γ_1 part of the boundary. Hence when the point 'i' becomes a boundary point on a smooth boundary we can write,

$$\frac{1}{2}u_l^i + \int_{\Gamma_2} u_k p_{lk}^* \mathrm{d}\Gamma + \int_{\Gamma_1} \bar{u}_k p_{lk}^* \mathrm{d}\Gamma$$

$$= \int_{\Gamma_1} p_k u_{lk}^* \mathrm{d}\Gamma + \int_{\Gamma_2} \bar{p}_k u_{lk}^* \mathrm{d}\Gamma + \int_{\Omega} b_k u_{lk}^* \mathrm{d}\Omega \tag{4.66}$$

For the case of non-smooth boundaries the evaluation of the integral on Γ_ϵ is more difficult and a result different from $-½$ is obtained. Fortunately, explicit calculation of this value is not usually necessary as it can be obtained using the rigid body motions.

In general we will have a c^i coefficient for a point on a non-smooth boundary, such that Equation (4.66) can be written

$$c^i u_l^i + \int_\Gamma u_k p_{lk}^* \mathrm{d}\Gamma = \int_\Gamma p_k u_{lk}^* \mathrm{d}\Gamma + \int_\Omega b_k u_{lk}^* \mathrm{d}\Omega \tag{4.67}$$

Boundary elements

It is now more convenient to work with matrices than to carry on with the indicial notation. We can define the following

$$\mathbf{u}^i = \begin{Bmatrix} u_1 \\ u_2 \\ u_3 \end{Bmatrix} =$$ displacement vector at point 'i' with components in $x_1\ x_2\ x_3$ directions

$\mathbf{u}$ = displacement vector at any point on boundary Γ

$$\mathbf{p} = \begin{Bmatrix} p_1 \\ p_2 \\ p_3 \end{Bmatrix} =$$ tractions at any point on boundary Γ

$$\mathbf{b} = \begin{Bmatrix} b_1 \\ b_2 \\ b_3 \end{Bmatrix} =$$ body forces at any point on domain Ω

$$\mathbf{p}^* = \begin{bmatrix} p_{11}^* & p_{12}^* & p_{13}^* \\ p_{21}^* & p_{22}^* & p_{23}^* \\ p_{31}^* & p_{32}^* & p_{33}^* \end{bmatrix}$$ = matrix whose coefficients, p_{lk}^*, are the forces in k direction due to a unit force at 'i' acting in the 'l' direction

$$\mathbf{u}^* = \begin{bmatrix} u_{11}^* & u_{12}^* & u_{13}^* \\ u_{21}^* & u_{22}^* & u_{23}^* \\ u_{31}^* & u_{32}^* & u_{33}^* \end{bmatrix}$$ = matrix whose coefficients u_{lk}^* are the displacements in the 'k' direction due to a unit force at 'i' acting in the 'l' direction

Equation (4.67) can now be expressed in matrix form as follows:

$$c^i\mathbf{u}^i + \int_\Gamma \mathbf{p}^*\mathbf{u}\,d\Gamma = \int_\Gamma \mathbf{u}^*\mathbf{p}\,d\Gamma + \int_\Omega \mathbf{u}^*\mathbf{b}\,d\Omega \tag{4.68}$$

We can assume that the boundary is divided into elements (Figure 4.6) and that the u and p functions can be approximated on each element 'j' using the following interpolation functions,

$$\mathbf{u} = \mathbf{\Phi}^T\mathbf{u}_j = \begin{bmatrix} \phi^T & . & . \\ . & \phi^T & . \\ . & . & \phi^T \end{bmatrix} \mathbf{u}_j$$

$$\mathbf{p} = \mathbf{\Phi}^T\mathbf{p}_j = \begin{bmatrix} \phi^T & . & . \\ . & \phi^T & . \\ . & . & \phi^T \end{bmatrix} \mathbf{p}_j \tag{4.69}$$

where $\mathbf{u}_j$ and $\mathbf{p}_j$ are the element nodal. Note that we have assumed the same functions for **u** and **p**. In general they may not be the same and it may be more consistent to take the functions for **p** of one order less than those for **u**.

The $\mathbf{\Phi}$ functions can be considered as the standard two dimensional finite element type functions[2,3]. $\mathbf{u}_j$ and $\mathbf{p}_j$ are the nodal displacements and tractions for an element and are unknown.

We can now substitute these functions into Equation (4.68) to obtain for a particular nodal point,

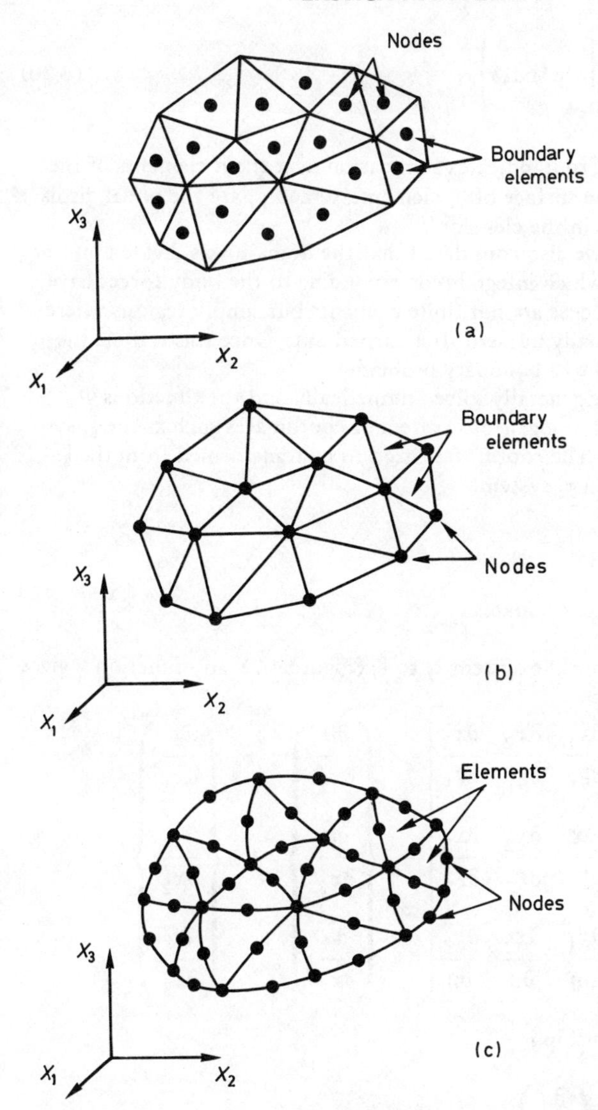

Fig.4.6. Three-dimensional body divided into (a) constant boundary elements, (b) linear boundary elements and (c) quadratic boundary elements

$$c^i \mathbf{u}^i + \sum_{j=1}^{n} \left\{ \int_{\Gamma_j} \mathbf{p}^* \mathbf{\Phi}^T d\Gamma \right\} \mathbf{u}_j = \sum_{j=1}^{n} \left\{ \int_{\Gamma_j} \mathbf{u}^* \mathbf{\Phi}^T d\Gamma \right\} \mathbf{p}_j +$$

$$+\sum_{s=1}^{m}\left\{\int_{\Omega_s} \mathbf{u}^*\mathbf{b}\,d\Omega\right\} \tag{4.70}$$

where Σ from $j = 1$ to n indicates summation over the n elements of the surface and Γ_j is the surface of 'j' element. $\mathbf{u}_j$ and $\mathbf{p}_j$ are the nodal displacement and tractions in the element 'j'.

Note that we have also considered that the domain was divided into m internal cells over which integrals corresponding to the body forces have to be computed. These are not finite elements but simply regions where the integration (usually numerical) is carried out. Once this is done the problem is reduced to a boundary problem.

The integrals are generally solved numerically and the functions Φ expressed in some homogeneous system of coordinates such as the ξ_i system of Figure 4.7. The coordinates need to be transformed from the ξ_i system to the global x_i system.

Transformation of coordinates

If we transform from the system x_i to ξ_i (Figure 4.7), any function u gives

$$\begin{Bmatrix} \dfrac{\partial u}{\partial \xi_1} \\ \dfrac{\partial u}{\partial \xi_2} \\ \dfrac{\partial u}{\partial \eta} \end{Bmatrix} = \begin{bmatrix} \dfrac{\partial x_1}{\partial \xi_1} & \dfrac{\partial x_2}{\partial \xi_1} & \dfrac{\partial x_3}{\partial \xi_1} \\ \dfrac{\partial x}{\partial \xi_2} & \dfrac{\partial x_2}{\partial \xi_2} & \dfrac{\partial x_3}{\partial \xi_2} \\ \dfrac{\partial x_1}{\partial \eta} & \dfrac{\partial x_2}{\partial \eta} & \dfrac{\partial x_3}{\partial \eta} \end{bmatrix} \begin{Bmatrix} \dfrac{\partial u}{\partial x_1} \\ \dfrac{\partial u}{\partial x_2} \\ \dfrac{\partial u}{\partial x_3} \end{Bmatrix} = \mathbf{J} \begin{Bmatrix} \dfrac{\partial u}{\partial x_1} \\ \dfrac{\partial u}{\partial x_2} \\ \dfrac{\partial u}{\partial x_3} \end{Bmatrix}$$

The inverse relationship is

$$\begin{Bmatrix} \dfrac{\partial u}{\partial x_1} \\ \dfrac{\partial u}{\partial x_2} \\ \dfrac{\partial u}{\partial x_3} \end{Bmatrix} = \mathbf{J}^{-1} \begin{Bmatrix} \dfrac{\partial u}{\partial \xi_1} \\ \dfrac{\partial u}{\partial \xi_2} \\ \dfrac{\partial u}{\partial \eta} \end{Bmatrix}$$

A differential of volume can be written,

$$\text{d (Volume)} = d\Omega = \text{Magnitude of} \left(\frac{\partial \vec{r}}{\partial \xi_1} \mathbf{x} \frac{\partial \vec{r}}{\partial \xi_2} \cdot \frac{\partial \vec{r}}{\partial \eta} \right) d\xi_1 d\xi_2 d\eta$$

$$= (\text{Absolute value of } |\mathbf{J}|) d\xi_1 d\xi_2 d\eta$$

A differential of area ($\eta = 0$ in Figure 4.7) will be given by

$$\text{d(Area)} = \left| \frac{\partial \vec{r}}{\partial \xi_1} \mathbf{x} \frac{\partial \vec{r}}{\partial \xi_2} \right| d\xi_1 d\xi_2 = |G| d\xi_1 d\xi_2 \tag{4.71}$$

Where

$$|G| = (g_1^2 + g_2^2 + g_3^2)^{1/2}$$

$$g_1 = \left(\frac{\partial x_2}{\partial \xi_1} \frac{\partial x_3}{\partial \xi_2} - \frac{\partial x_2}{\partial \xi_2} \frac{\partial x_3}{\partial \xi_1} \right), \quad g_2 = \left(\frac{\partial x_1}{\partial \xi_1} \frac{\partial x_3}{\partial \xi_2} - \frac{\partial x_1}{\partial \xi_2} \frac{\partial x_3}{\partial \xi_1} \right)$$

$$g_3 = \left(\frac{\partial x_1}{\partial \xi_1} \frac{\partial x_2}{\partial \xi_2} - \frac{\partial x_1}{\partial \xi_2} \frac{\partial x_2}{\partial \xi_1} \right)$$

Note that $x_1 x_2 x_3$ can be defined using interpolation functions and nodal values in the same way as the displacements and tractions were defined (Equation 4.69).

Applying numerical integration (see Appendix) Equation (4.70 becomes

$$c^i \mathbf{u}^i + \sum_{j=1}^{n} \left\{ \sum_{k=1}^{l} |G| w_k (\mathbf{p}^* \mathbf{\Phi}^T)_k \right\} u_j = \sum_{j=1}^{n} \left\{ \sum_{k=1}^{l} |G| w_k (\mathbf{u}^* \mathbf{\Phi}^T)_k \right\} \mathbf{p}_j$$

$$+ \sum_{s=1}^{m} \left\{ \sum_{p=1}^{n} |\mathbf{J}| w_p (u^* b)_p \right\} \tag{4.72}$$

where l = number of integration points, w = weight at the integration points, $(\mathbf{p}^* \mathbf{\Phi}^T)_k$, $(\mathbf{u}^* \mathbf{\Phi}^T)_k$ and $(\mathbf{u}^* \mathbf{b})_p$ are the values of the functions at the integration points.

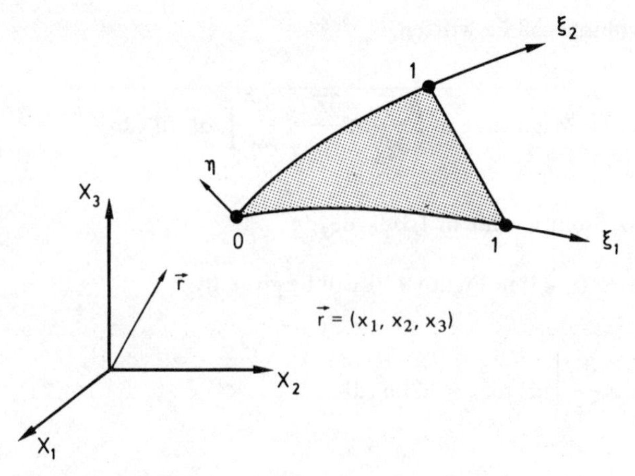

Fig.4.7. Co-ordinate system for a boundary element

System of equations

Equation (4.72) corresponds to a particular node '*i*'. The terms integrated numerically relate the '*i*' node with the nodes of element '*j*' over which the integration is carried out. We call the left hand side integral $\hat{\mathbf{H}}_{ij}$ and the right hand side $\mathbf{G}_{ij}$.

The *H* and *G* contributions can be assembled together in a global matrix such that,

$$\mathbf{C\,U} + \hat{\mathbf{H}}\,\mathbf{U} = \mathbf{G\,P} + \mathbf{B} \tag{4.73}$$

or

$$\mathbf{H\,U} = \mathbf{G\,P} + \mathbf{B} \tag{4.74}$$

where $\mathbf{H} = \mathbf{C} + \hat{\mathbf{H}}$, and $\mathbf{C}$ is a diagonal matrix, which does not need to be determined explicitly. $\mathbf{U}$ are the displacements and $\mathbf{P}$ the values that the distributed tractions take at the boundary nodes.

The diagonal coefficients in the *H* matrix can be obtained by applying rigid body conditions. If we assume a unit rigid body displacement in any direction, Equation (4.74) for a closed domain becomes,

$$\mathbf{H\,I} = \mathbf{O} \tag{4.75}$$

where $\mathbf{I}$ is a vector of unit displacement in the chosen direction. Hence all the terms in the diagonal sub-matrices of $\mathbf{H}$ can be obtained as in p. 58.

After applying the boundary conditions the system can be reordered and we obtain,

$$\mathbf{A\,X = F} \tag{4.76}$$

where **X** contains the unknown displacements and tractions.

Internal points

Once the boundary values are known we can compute the internal values of displacements and stresses. The displacements at a point are given by,

$$\mathbf{u}^i = \int_\Gamma \mathbf{u}^*\mathbf{p}\,d\Gamma - \int_\Gamma \mathbf{p}^*\mathbf{u}\,d\Gamma + \int_\Omega \mathbf{b}\mathbf{u}^*\,d\Omega \tag{4.77}$$

or for 'l' component

$$u_l^i = \int_\Gamma u_{lk}^* p_k\,d\Gamma - \int_\Gamma p_{lk}^* u_k\,d\Gamma + \int_\Omega b_k u_{lk}^*\,d\Omega \tag{4.78}$$

For an isotropic medium the stresses can now be calculated by differentiating **u** at the internal points, i.e.

$$\sigma_{ij} = \frac{2G\nu}{1-2\nu}\,\Delta_{ij}\,\frac{\partial u_l}{\partial x_l} + G\left(\frac{\partial u_i}{\partial x_j} + \frac{\partial u_j}{\partial x_i}\right) \tag{4.79}$$

After carrying out the differentiation we can obtain,

$$\begin{aligned}
\sigma_{ij} = &\int_\Gamma \left\{\frac{2G\nu}{1-2\nu}\,\Delta_{ij}\,\frac{\partial u_{lk}^*}{\partial x_l} + G\left(\frac{\partial u_{ik}^*}{\partial x_j} + \frac{\partial u_{jk}^*}{\partial x_i}\right)\right\} p_k\,d\Gamma \\
&+ \int_\Omega \left\{\frac{2G\nu}{1-2\nu}\,\Delta_{ij}\,\frac{\partial u_{lk}^*}{\partial x_l} + G\left(\frac{\partial u_{ik}^*}{\partial x_j} + \frac{\partial u_{jk}^*}{\partial x_i}\right)\right\} b_k\,d\Omega \\
&- \int_\Gamma \left\{\frac{2G\nu}{1-2\nu}\,\Delta_{ij}\,\frac{\partial p_{lk}^*}{\partial x_l} + G\left(\frac{\partial p_{ik}^*}{\partial x_j} + \frac{\partial p_{jk}^*}{\partial x_i}\right)\right\} u_k\,d\Gamma
\end{aligned} \tag{4.80}$$

All these derivatives are taken at the internal points. Taking into account the way the derivatives of r are calculated (Figure 4.8) we can reduce this expression to

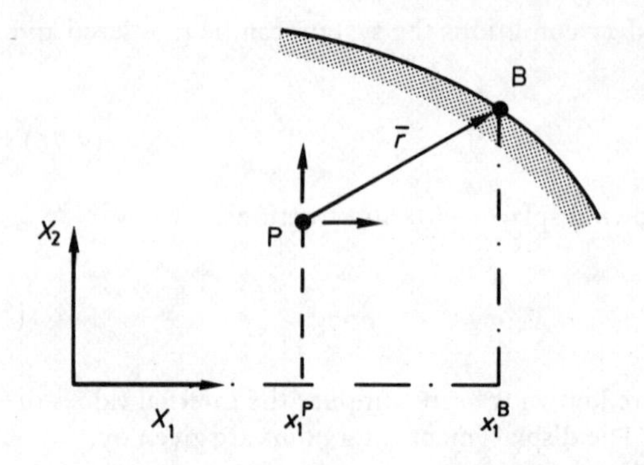

Fig.4.8. Derivatives definition. x_i^B: *boundary point co-ordinates in 'i' direction*
x_i^P: *point P coordinates in 'i' direction*

Note that $\partial r/\partial x_i^B = r_{,i} = (x_i^B - x_i^P)/r$

$$\sigma_{ij} = \int_\Gamma D_{kij} p_k d\Gamma - \int_\Gamma S_{kij} u_k d\Gamma + \int_\Omega D_{kij} b_k d\Omega \tag{4.81}$$

where the third order tensor components D_{kij} and S_{kij} are

$$D_{kij} = \frac{1}{r^\alpha}\{(1-2\nu)\{\Delta_{ki}r_{,j} + \Delta_{kj}r_{,i} - \Delta_{ij}r_{,k}\} + \beta r_{,i}r_{,j}r_{,k}\}\frac{1}{4\alpha\pi(1-\nu)} \tag{4.82}$$

$$S_{kij} = \frac{2\mu}{r^\beta}\left\{\beta\frac{\partial r}{\partial n}[(1-2\nu)\Delta_{ij}r_{,k} + \nu(\Delta_{ik}r_{,j} + \Delta_{jk}r_{,i}) - \gamma r_{,i}r_{,j}r_{,k}] + \beta\nu(n_i r_{,j}r_{,k} + n_j r_{,i}r_{,k}) + (1-2\nu)(\beta n_k r_{,i}r_{,j} + n_j\Delta_{ik} + n_i\Delta_{jk}) - (1-4\nu)n_k\Delta_{ij}\right\}\frac{1}{4\alpha\pi(1-\nu)} \tag{4.83}$$

where $r_{,i} = \partial r/\partial x_i$.
(The derivatives are taken on the boundary (Figure (4.8)).)

The above formulae apply for two and three dimensional cases, i.e.

(1) For two dimensions, $\alpha = 1$; $\beta = 2$; $\gamma = 4$
(2) For three dimensions, $\alpha = 2$; $\beta = 3$; $\gamma = 5$.

4.4 THREE-DIMENSIONAL APPLICATIONS

In order to illustrate the applications of the method for linear elastic problems we will discuss two cases presented in reference [4]. They are

(1) The analysis of a thick cylinder.
(2) The study of a pipe connection.

Case (1). Thick cylinder

This study was carried out to compare the results obtained using boundary elements with those due to the finite element method and the exact solution. Two types of boundary elements were used, i.e. linear and quadratic elements.

The dimensions of the cylinder are shown in Figure 4.9. The cylinder was subjected to an internal pressure of 20 N/mm^2. A 90° sector was analysed taking into account the symmetry. The finite element discretization shown in Figure 4.9(b) corresponds to 20-node isoparametric elements. Table 4.1 shows the calculated and exact results for five values of radius over the thickness of the cylinder. Special care was taken in concentrating the integration points around the singularities for the boundary element method in order to avoid numerical errors. It is interesting to note that the results obtained for the linear boundary element variation are less accurate than the finite element results but those obtained with quadratic variation are equally good for displacements and better for stresses than the finite element results.

Execution time for quadratic boundary elements and finite elements were approximately equal but the thick cylinder problem is a comparatively small one and not very representative.

Case (2). The pipe connection

This problem is that of a flange and has been the subject of extensive experimental and theoretical work. The flange connects two pipes carrying fluid under pressure, the arrangement is symmetric with respect to the plane of the joint and with respect to the 12 connecting bolts. Because of this only a 15° sector need to be considered and for the numerical analysis only 100 mm of pipe were taken into account. The dimensions of the section

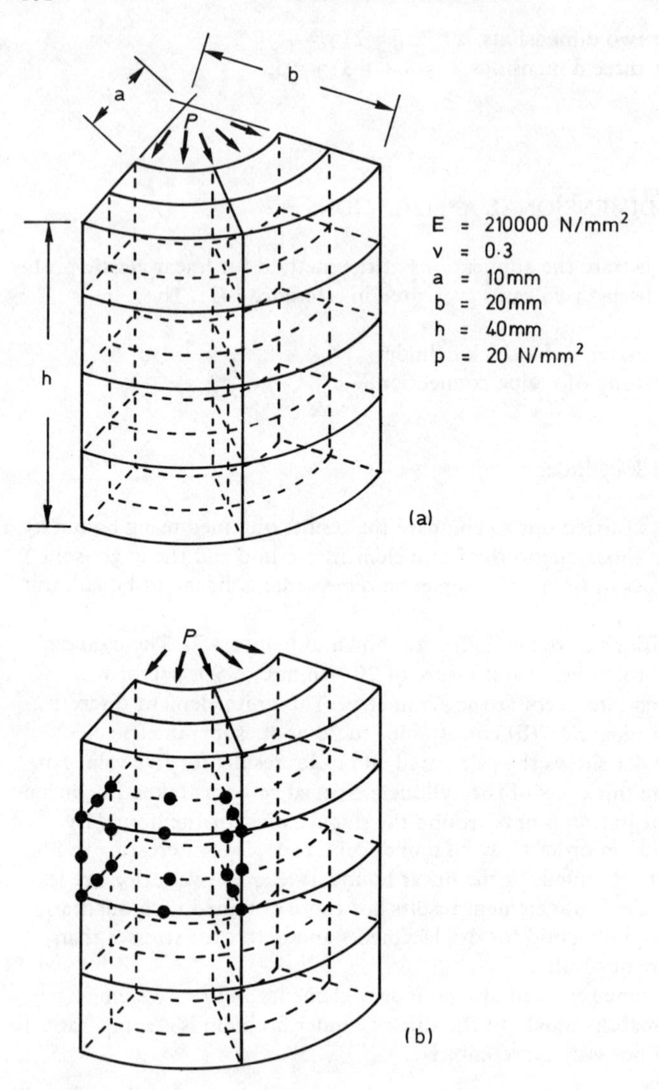

Fig.4.9. (a) Boundary element network and (b) finite element mesh of thick cylinder

analysed are shown in Figure 4.10. The load case reported here is with a bolt tension of 50000 N/bolt plus an internal pressure of 4.5 N/mm^2. (For other results see reference [4].)

The boundary element discretization consisted of 57 surface and interface elements as shown in Figure 4.11. The finite element mesh, consist-

Table 4.1. Comparison between calculated and exact solutions

Function	Radius	Exact solution	Finite element	Boundary elements	
				Linear	Quadratic
Radial displacement	10.0	1.904	1.905	1.818	1.905
	12.5	1.602	1.600	1.568	1.600
	15.0	1.415	1.414	1.319	1.414
	17.5	1.293	1.292	1.234	1.291
	20.0	1.212	1.212	1.150	1.211
Radial stress	10.0	−20.0	−17.4	−13.8	−18.5
	12.5	−10.4	−11.6	−13.2	−11.3
	15.0	−5.2	−3.5	−1.8	−4.1
	17.5	−2.0	−2.4	−1.2	−2.2
	20.0	0.0	0.7	1.1	0.4
Hoop stress	10.0	33.3	34.4	27.1	33.5
	12.5	23.7	23.3	20.1	23.7
	15.0	18.5	19.2	15.2	18.5
	17.5	15.4	15.2	14.3	15.4
	20.0	13.3	13.5	12.1	13.3

ing of 64 twenty-node isoparametric elements and 501 nodes, is shown in Figure 4.12.

The execution time for the analysis by the boundary element method was 58 seconds or approximately 65% greater than that for the finite element analysis. Figures 4.13(a), (b) and (c) show the variation of meridonal and circumferential stresses over the outer and inner surfaces for the load case under consideration.

All results show that there is a strong stress concentration in the radius between the plate and the cone and a weaker stress concentration at the junction between the cone and the pipe. The fact that the value obtained by the finite element method is higher than that given by the boundary element method at the junction is due to the fineness of the finite element mesh. By putting more finite or boundary elements near the junction better results for stresses could be obtained.

Lachat[4] concluded that the problem of the flange is relatively expensive to solve by the boundary element method because the surface area/volume ratio is high, and the finite element method is economical because the bandwidth is small. In the analysis by boundary elements, most of the computing time is spent integrating the elements of the matrix, whereas in the finite element method the reduction of the system of equations is usually the longest calculation.

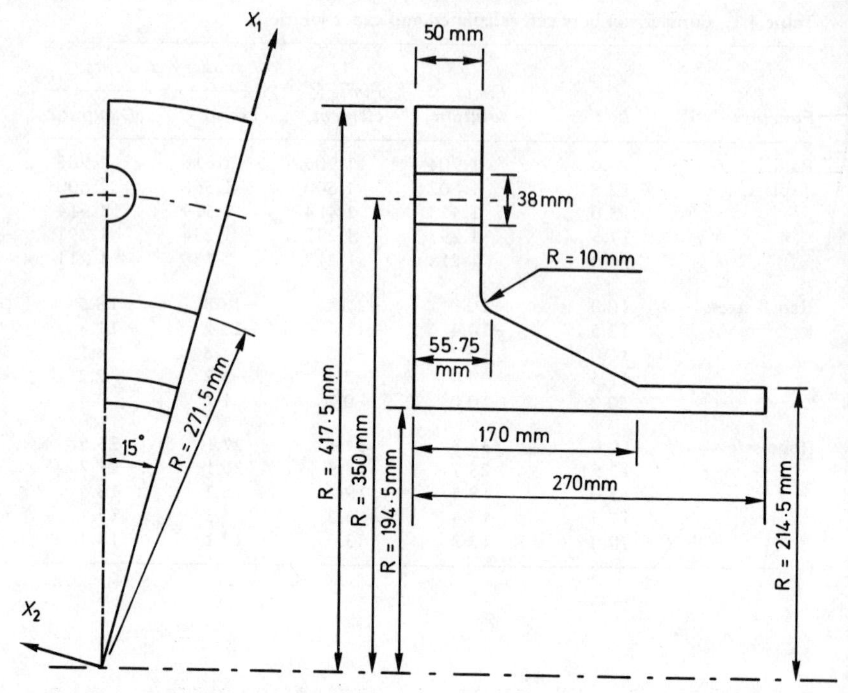

Fig.4.10. Dimensions of flange

4.5 INITIAL STRESS OR STRAIN FIELDS

It is easy to show that initial stress or strain fields can be treated in a similar way as the body forces field.

Consider the starting Equation (4.47) i.e.

$$\int_{\Omega} (\sigma_{jk,j} + b_k)u_k^* d\Omega = \int_{\Gamma_2} (p_k - \bar{p}_k)u_k^* d\Gamma + \int_{\Gamma_1} (\bar{u}_k - u_k)p_k^* d\Gamma \qquad (4.84)$$

plus the strain-displacements relationships (4.48) and the stress–strain Equations (4.18) with initial strains, i.e.

$$\sigma_{ij} = \sigma_{ij}^t + \sigma_{ij}^0 \qquad (4.85)$$

where σ_{ij}^t represents the 'total' stress and σ_{ij}^0 the initial one.

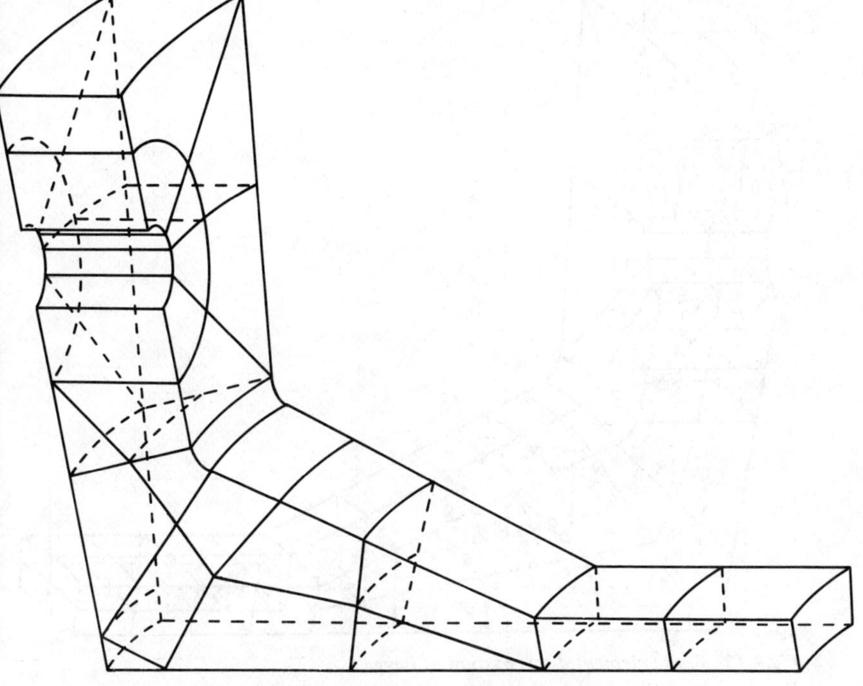

Fig.4.11. Boundary element discretization of surface of flange

Integrating Equation (4.84) by parts once, gives the following relationship:

$$\int_{\Omega} b_k u_k^* \mathrm{d}\Omega - \int_{\Omega} \sigma_{jk}\epsilon_{jk}^* \mathrm{d}\Omega = -\int_{\Gamma_2} \bar{p}_k u_k^* \mathrm{d}\Gamma - \int_{\Gamma_1} p_k u_k^* \mathrm{d}\Gamma$$

$$+ \int_{\Gamma_1} (\bar{u}_k - u_k) p_k^* \mathrm{d}\Gamma \qquad (4.86)$$

Substituting Equation (4.85) we can write,

$$\int_{\Omega} b_k u_k^* \mathrm{d}\Omega - \int_{\Omega} \sigma_{jk}^t \epsilon_{jk}^* \mathrm{d}\Omega - \int_{\Omega} \sigma_{jk}^0 \epsilon_{jk}^* \mathrm{d}\Omega$$

$$= -\int_{\Gamma_2} \bar{p}_k u_k^* \mathrm{d}\Gamma - \int_{\Gamma_1} p_k u_k^* \mathrm{d}\Gamma + \int_{\Gamma_1} (\bar{u}_k - u_k) p_k^* \mathrm{d}\Gamma \qquad (4.87)$$

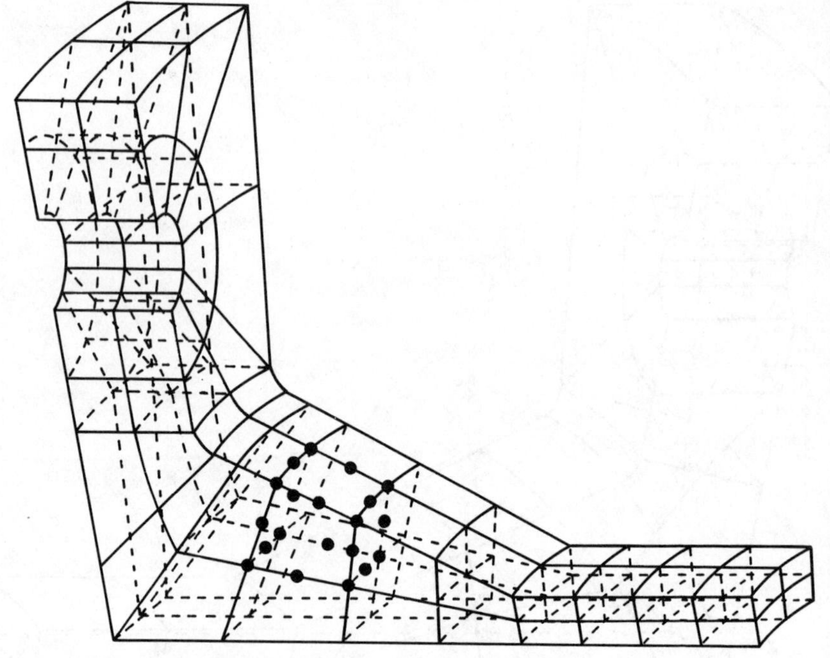

Fig.4.12. Finite element discretization of flange

Integrating by parts, once more, we obtain,

$$\int_{\Omega} b_k u_k^* d\Omega + \int_{\Omega} \sigma_{jk,j}^* u_k d\Omega - \int_{\Omega} \sigma_{jk}^0 \epsilon_{jk}^* d\Omega$$

$$= -\int_{\Gamma_2} \bar{p}_k u_k^* d\Gamma - \int_{\Gamma_1} p_k u_k^* d\Gamma + \int_{\Gamma_1} \bar{u}_k p_k^* d\Gamma + \int_{\Gamma_2} u_k p_k^* d\Gamma \qquad (4.88)$$

This shows that any initial stress (or strain) field can be treated in a similar way to the body force field, b_k. In addition to being used in temperature and other problems the initial fields are important because their use allows non-linearities to be introduced in the formulations.

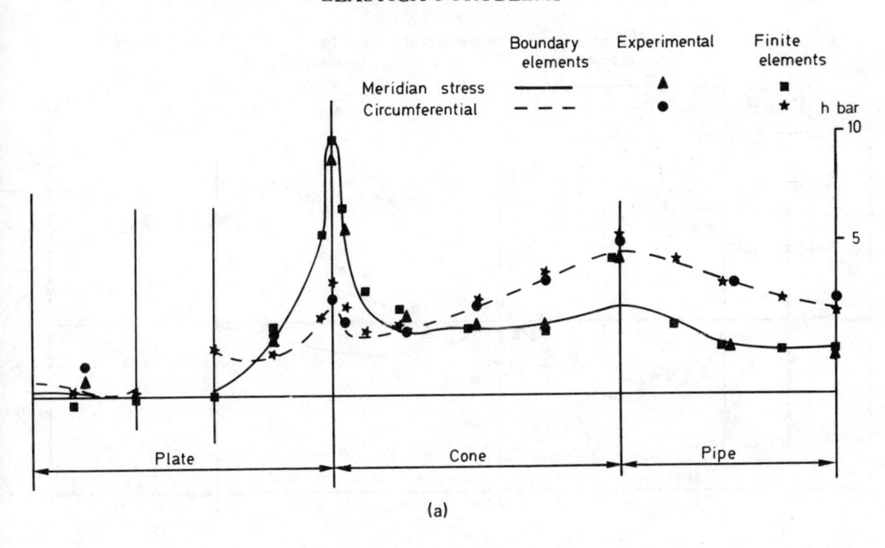

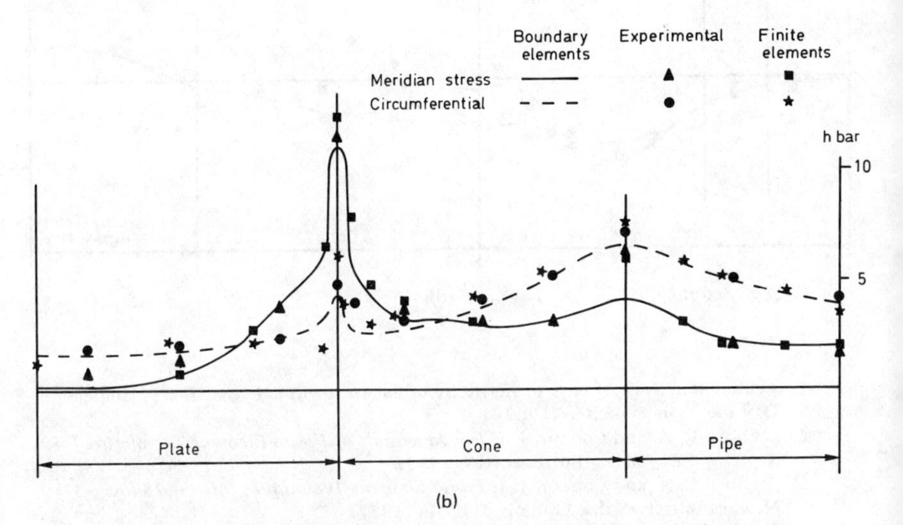

Fig.4.13. Plane of symmetry passing between two holes; (a) and (b) results on outer surface and (c) and (d) results on inner surface

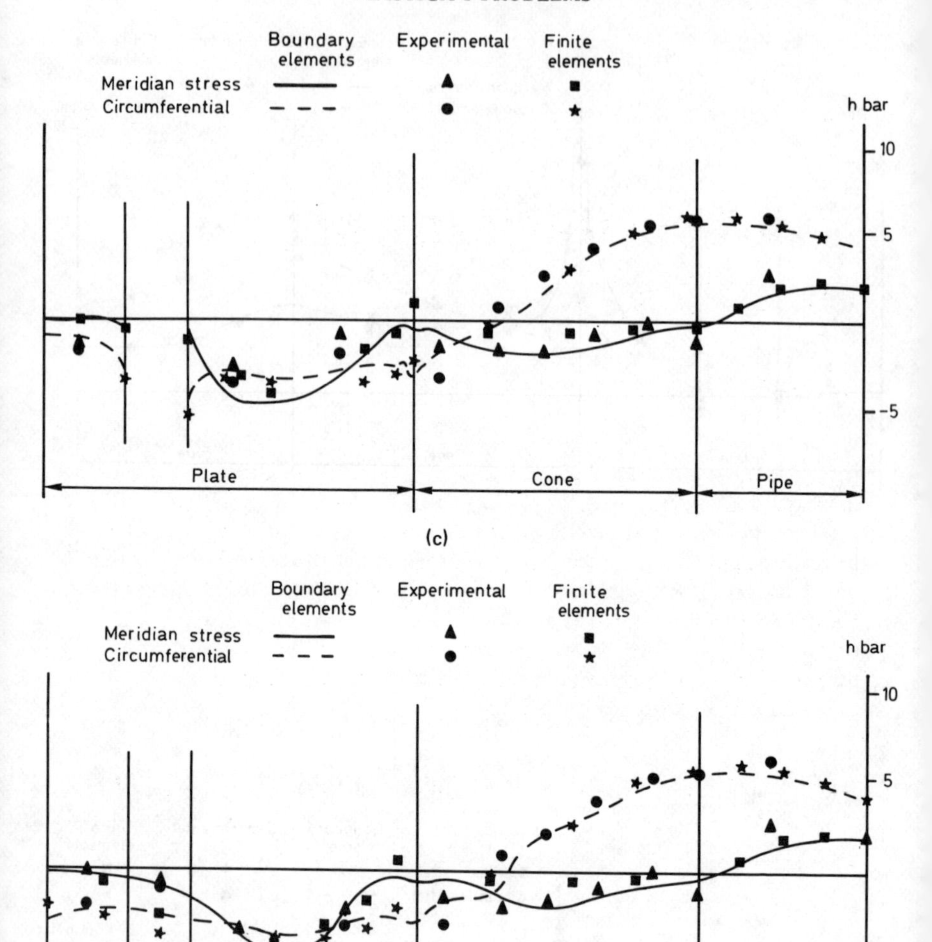

Fig.4.13 (contd.) (d)

References

1. Lekhnitskii, S. G. *Theory of Elasticity of an Anisotropic Elastic Body*, Holden-Day Inc., San Francisco (1963)
2. Brebbia, C. A. and Connor, J. J. *Fundamentals of Finite Element Techniques for Structural Engineers*, Butterworths (1973)
3. Brebbia, C. A. and Connor, J. J. *Finite Element Techniques for Fluid Flow*, Newnes-Butterworths, London, 2nd edn (1977)
4. Lachat, J. C., *A Further Development of the Boundary Integral Technique for Elastostatics*, Ph.D. Thesis, Southampton University (1975)

5

Two-dimensional elasticity

5.1 INTRODUCTION

We will now apply the boundary element method in two-dimensional elasticity. The basic relationships and governing equations have been presented in Chapter 4, together with the fundamental solution for an isotropic body.

A computer program will now be developed to show how the theory can be readily applied in engineering. The program has the same format as that of the potential programs developed in Chapter 3. We will only discuss the constant element program as it seems to be very appropriate for practical applications. With this basis the reader can build his own computer program for linear, quadratic or even higher order boundary elements.

We start with Equation (4.67), i.e.

$$c^i u_l^i + \int_\Gamma u_k p_{lk}^* \mathrm{d}\Gamma = \int_\Gamma p_k u_{lk}^* \mathrm{d}\Gamma + \int_\Omega b_k u_{lk}^* \mathrm{d}\Omega \tag{5.1}$$

where the fundamental solution is assumed to satisfy,

$$\sigma_{jk,j}^* + \Delta_l^i = 0 \tag{5.2}$$

Equation (5.2) applies for a point in the boundary and if $c = 1$ it also applies for an internal point.

The fundamental solution for an isotropic material has been given in Chapter 4, Equation (4.57), but it is repeated below for completeness.

$$u_{lk}^* = \frac{1}{8\pi G(1-\nu)}\left[(3-4\nu)\ln\left(\frac{1}{r}\right)\Delta_{lk} + \frac{\partial r}{\partial x_l}\frac{\partial r}{\partial x_k}\right]$$

$$p_{lk}^* = -\frac{1}{4\pi(1-\nu)r}\left[\frac{\partial r}{\partial n}\left\{(1-2\nu)\Delta_{kl} + 2\frac{\partial r}{\partial x_k}\frac{\partial r}{\partial x_l}\right\} - (1-2\nu)\left(\frac{\partial r}{\partial x_l}n_k - \frac{\partial r}{\partial x_k}n_l\right)\right] \tag{5.3}$$

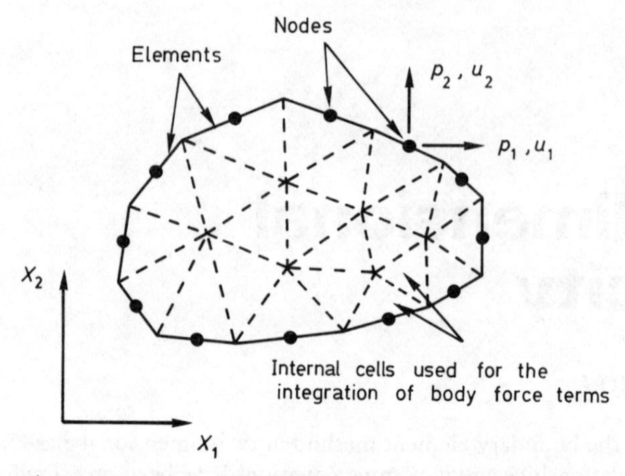

Fig. 5.1. Two-dimensional body divided into boundary elements and internal cells

where p^*_{lk} and u^*_{lk} represent the tractions and displacements in the k direction due to a unit force in the direction l.

In matrix form $\mathbf{u}^*$ is a 2 × 2 matrix with elements u^*_{lk} and $\mathbf{p}^*$ is also a (2 × 2) matrix with elements p^*_{lk}, i.e.

$$\mathbf{p}^* = \begin{bmatrix} p^*_{11} & p^*_{12} \\ p^*_{21} & p^*_{22} \end{bmatrix} ; \qquad \mathbf{u}^* = \begin{bmatrix} u^*_{11} & u^*_{12} \\ u^*_{21} & u^*_{22} \end{bmatrix} \tag{5.4}$$

The displacements and tractions and the known body forces are

$$\mathbf{u} = \begin{Bmatrix} u_1 \\ u_2 \end{Bmatrix}, \qquad \mathbf{p} = \begin{Bmatrix} p_1 \\ p_2 \end{Bmatrix}, \qquad \mathbf{b} = \begin{Bmatrix} b_1 \\ b_2 \end{Bmatrix}, \tag{5.5}$$

The basic boundary equation can be written as,

$$c^i\mathbf{u}^i + \int_\Gamma \mathbf{p}^*\mathbf{u}\,d\Gamma = \int_\Gamma \mathbf{u}^*\mathbf{p}\,d\Gamma + \int_\Omega \mathbf{u}^*\mathbf{b}\,d\Omega \tag{5.6}$$

The c^i constants are generally determined from the rigid body conditions as indicated in Chapter 4.

Let us now consider the case of constant elements (Figure 5.1) each of them with a node at the middle. The values of $\mathbf{p}$ and $\mathbf{u}$ are assumed to be constant on each element and equal to the value at the mid-node of the element. Hence Equation (5.6) becomes,

$$c^i \mathbf{u}^i + \sum_{j=1}^{n} \left\{ \int_{\Gamma_j} \mathbf{p}^* d\Gamma \right\} \mathbf{u}_j = \sum_{j=1}^{n} \left\{ \int_{\Gamma_j} \mathbf{u}^* d\Gamma \right\} \mathbf{p}_j + \int_{\Omega} \mathbf{u}^* \mathbf{b} d\Omega \tag{5.7}$$

where $\mathbf{u}_j$ and $\mathbf{p}_j$ are the nodal displacement and traction in the element 'j'. Note that in addition it is convenient to define internal cells for integration of the body forces. These cells are used only for the numerical integration of the body force terms and should not be confused with finite elements. If there are m of these cells we can write

$$\int_{\Omega} \mathbf{u}^* \mathbf{b} d\Omega = \sum_{s=1}^{m} \left\{ \sum_{p=1}^{l} (\mathbf{u}^* \mathbf{b})_p w_p \right\} A_s = \mathbf{b}^i \tag{5.8}$$

where w_p are the weighting coefficients for the numerical integration and A_s is the area of the element under consideration. Evaluation of the body force terms produces a vector $\mathbf{b}$.

Equation (5.7) corresponds to a particular node 'i'. The terms

$$\int_{\Gamma_j} \mathbf{p}^* d\Gamma \text{ and } \int_{\Gamma_j} \mathbf{u}^* d\Gamma$$

relate the 'i' node with the segment 'j' over which the integral is carried out. We call these integrals $\hat{\mathbf{H}}_{ij}$ and $\mathbf{G}_{ij}$ and they are now 2×2 matrices. Hence one has,

$$\mathbf{c}^i \mathbf{u}^i + \sum_{j=1}^{n} \hat{\mathbf{H}}_{ij} \mathbf{u}_j = \sum_{j=1}^{n} \mathbf{G}_{ij} \mathbf{p}_j + \mathbf{b}^i \tag{5.9}$$

This equation relates the value of u at mid-node 'i' with the values of u's and p's at all the nodes on the boundary, including 'i'.

One can write Equation (5.9) for each 'i' node obtaining $2 \times n$ equations where n is the total number of nodes. Let us now call

$$\mathbf{H}_{ij} = \hat{\mathbf{H}}_{ij}, \text{ when } i \neq j$$

$$\mathbf{H}_{ij} = \hat{\mathbf{H}}_{ij} + \mathbf{c}^i, \text{ when } i = j \tag{5.10}$$

where $\mathbf{c}^i$ is a coefficient matrix depending on the boundary geometry, i.e.

$$\mathbf{c}^i = \begin{bmatrix} c^i & 0 \\ 0 & c^i \end{bmatrix} \tag{5.11}$$

with $c^i = ½$ for smooth boundaries.

Hence Equation (5.9) can be written as,

$$\sum_{j=1}^{n} \mathbf{H}_{ij}\mathbf{u}_j = \sum_{j=1}^{n} \mathbf{G}_{ij}\mathbf{p}_j + \mathbf{b}^i \tag{5.12}$$

The whole set of equations for the n boundary nodes can be expressed in matrix form as,

$$\mathbf{H\,U} = \mathbf{G\,P} + \mathbf{B} \tag{5.13}$$

Note that n_1 values of displacements and n_2 values of tractions are known ($2n = n_1 + n_2$), hence one has a set of $2n$ unknowns in Equation (5.13). Re-ordering the equations in the same way as for the potential problems, i.e. with the unknowns on the left hand side vector **X**, we obtain,

$$\mathbf{A\,X} = \mathbf{F} + \mathbf{B} \tag{5.14}$$

Internal points

Once the values of displacements and tractions are known on the boundary one can calculate the displacements and stresses at any interior point, i.e.

$$\mathbf{u}^i = \int_\Gamma \mathbf{u}^*\mathbf{p}\,d\Gamma - \int_\Gamma \mathbf{p}^*\mathbf{u}\,d\Gamma + \int_\Omega \mathbf{u}^*\mathbf{b}\,d\Omega$$

$$\mathbf{u}^i = \sum_{k=1}^{n} \left\{ \int_{\Gamma_k} \mathbf{u}^*\,d\Gamma \right\} \mathbf{p}_k - \sum_{k=1}^{n} \left\{ \int_{\Gamma_k} \mathbf{p}^*\,d\Gamma \right\} \mathbf{u}_k$$

$$+ \sum_{s=1}^{m} \left\{ \sum_{p=1}^{l} (\mathbf{u}^*\mathbf{b})_k w_k \right\} A_s \tag{5.15}$$

$$\sigma_{ij} = \int_\Gamma \mathbf{D}_{ij}\mathbf{p}\,d\Gamma - \int_\Gamma \mathbf{S}_{ij}\mathbf{u}\,d\Gamma + \int_\Omega \mathbf{D}_{ij}\mathbf{b}\,d\Omega$$

$$\sigma_{ij} = \sum_{k=1}^{n} \left\{ \int_{\Gamma_k} \mathbf{D}_{ij}\,d\Gamma \right\} \mathbf{p}_k - \sum_{k=1}^{n} \left\{ \int_{\Gamma_k} \mathbf{S}_{ij}\,d\Gamma \right\} \mathbf{u}_k + \sum_{s=1}^{m} \left\{ \int_{\Omega_s} \mathbf{D}_{ij}\,d\Omega \right\} \mathbf{b}_s \tag{5.16}$$

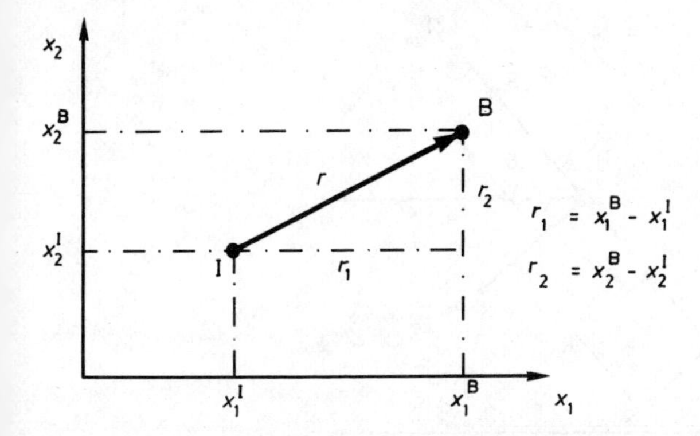

Fig. 5.2. Geometrical definitions for the points I and B

where

$$\mathbf{D}_{ij} = [D_{1ij} \quad D_{2ij}], \qquad \mathrm{p} = \begin{Bmatrix} p_1 \\ p_2 \end{Bmatrix}$$

$$\mathbf{S}_{ij} = [S_{1ij} \quad S_{2ij}], \qquad \mathrm{u} = \begin{Bmatrix} u_1 \\ u_2 \end{Bmatrix} \tag{5.17}$$

The values of D and S have been given in Equations (4.82) and (4.83) for the two-dimensional case.

Integration

The integrals for $\mathbf{H}_{ij}$ and $\mathbf{G}_{ij}$ have been calculated everywhere using the four-point Gauss quadrature formulae, except for the case $i = j$. The values of $\mathbf{H}_{ii}$ are simple to calculate using rigid body considerations but $\mathbf{G}_{ii}$ would need to be computed using a logarithmically weighted numerical integration formula (see Appendix 1).

Although numerical integration is generally more convenient it is not difficult to calculate $\mathbf{G}_{ii}$ analytically for the isotropic two-dimensional case. The elements of $\mathbf{G}_{ii}$ are

$$\mathbf{G}_{ii} = \begin{bmatrix} G_{11} & G_{12} \\ G_{21} & G_{22} \end{bmatrix}_{ii} \tag{5.18}$$

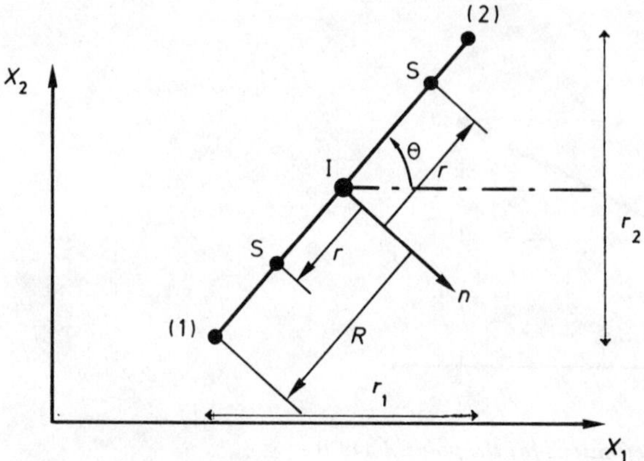

Fig. 5.3. Element definition

With the definitions shown in Figures 5.2 and 5.3 the elements of $\mathbf{G}_{ii}$ are,

$$G_{11} = \left[(3-4\nu)\int_{(1)}^{(2)} \ln\left(\frac{1}{r}\right) d\Gamma + \int_{(1)}^{(2)} \left(\frac{\partial r}{\partial x_1}\right)^2 d\Gamma\right] \frac{1}{8\pi G(1-\nu)}$$

$$= \lim_{\epsilon\to 0}\left[2(3-4\nu)\int_{\epsilon}^{R} \ln\frac{1}{r}\, dr + 2\int_{\epsilon}^{R} \cos^2\theta\, dr\right] \frac{1}{8\pi G(1-\nu)}$$

$$= \lim_{\epsilon\to 0}\left[2(3-4\nu)\left\{r\ln\left(\frac{1}{r}\right)\Bigg|_{\epsilon}^{R} + r\Bigg|_{\epsilon}^{R}\right\} + 2R\cos^2\theta\right] \frac{1}{8\pi G(1-\nu)}$$

$$= \frac{2R}{8\pi G(1-\nu)}\left[(3-4\nu)(1-\ln R) + \frac{(r_1)^2}{4R^2}\right]$$

$$G_{12} = \left[\int_{(1)}^{(2)} \frac{\partial r}{\partial x_1}\frac{\partial r}{\partial x_2}\, d\Gamma\right] \frac{1}{8\pi G(1-\nu)}$$

$$= \lim_{\epsilon\to 0}\left[\int_{0}^{R} \sin\theta\cos\theta\, dr + \int_{0}^{R} \sin\theta\cos\theta\, dr\right] \frac{1}{8\pi G(1-\nu)}$$

$$= [2R \sin\theta \cos\theta] \frac{1}{8\pi G(1-\nu)} = \left[2R \frac{r_1 r_2}{4R^2}\right] \frac{1}{8\pi G(1-\nu)} = G_{21}$$

$$G_{22} = \left[(3-4\nu) \int_{(1)}^{(2)} \ln \frac{1}{r} \, d\Gamma + \int_{(1)}^{(2)} \left(\frac{\partial r}{\partial x_2}\right)^2 d\Gamma\right] \frac{1}{8\pi G(1-\nu)}$$

$$= [2(3-4\nu)R(1-\ln R) + 2R\sin^2\theta] \frac{1}{8\pi G(1-\nu)}$$

$$= \left[(3-4\nu)(1-\ln R) + \frac{(r_2)^2}{4R^2}\right] \frac{2R}{8\pi G(1-\nu)} \qquad (5.19)$$

The definitions for r_1 and r_2 are given in Figure 5.2.

5.2 COMPUTER PROGRAM

In what follows a FORTRAN computer program for the solution of isotropic linearly elastic problems without body forces is described. The inclusion of body forces does not present any special problems and may be of interest for the solution of non-linear problems but it has not been attempted here for simplicity.

The program uses 'constant' elements and has the same organization as the one described in Section 3.4.

It is important to note that constant elements, although accurate for many practical applications should be used with caution when the problem involves bending and rigid body components.

Main program and data structure

The general variables and arrays used in the program together with their meaning are as follows:

- N: Number of boundary elements (equal to number of nodes for this case).
- L: Number of internal points where the displacements and stresses are to be calculated.
- M: Number of different surfaces (maximum 5).
- NC ( ): Number of the last node of each different surface.
- GE: Shear modulus.
- XNU: Poisson's ratio.

X, Y: One dimensional arrays with x_1 and x_2 coordinates of the extreme points of the boundary elements.

XM, YM: One dimensional arrays with the coordinates of the nodes.

G: Matrix defined in Equation (5.13). After application of the boundary conditions the matrix **A** is stored in the same location (Equation 5.14).

H: Matrix defined by Equation (5.13).

KODE: One dimensional array indicating the type of boundary condition at the element nodes. KODE = 0 means that a displacement is prescribed and KODE = 1 that a traction is prescribed.

FI: Vector where the prescribed values of boundary conditions are stored. Each element is associated with a value of KODE.

DFI: Right hand side vector in the global system. After solution it contains the values of the unknowns.

CX, CY: One dimensional vectors with coordinates of internal points (maximum number 20).

DSOL: Values of displacements at internal points (2 displacements per point).

SSOL: Values of stresses at internal points (3 stresses per point).

Program 18

```
C
C  PROGRAM 18
C
C  PROGRAM FOR SOLUTION OF TWO DIMENSIONAL ELASTIC PROBLEMS
C  BY THE B. E. METHOD WITH CONSTANT ELEMENTS
C
      COMMON N, L, NC(5), M, GE, XNU, LEC, IMP
      DIMENSION X(59), Y(59), XM(58), YM(58), G(116, 116), FI (116), DFI(116)
      DIMENSION KODE (116), CX(20), CY(20), SSOL(60), DSOL(40)
      DIMENSION H(116, 116)
C
C  INITIALIZATION OF PROGRAM PARAMETERS
C  NX = MAXIMUM DIMENSION OF THE SYSTEM OF EQUATIONS.
C
      NX = 116
C
C  ASSIGN DATA SET NUMBERS FOR INPUT, LEC, AND OUTPUT, IMP
C
      LEC = 5
      IMP = 6
C
C  INPUT
C
      CALL INPUT (CX, CY, X, Y, KODE, FI)
C
C  FORM SYSTEM OF EQUATIONS
C
```

```
      CALL FMAT(X, Y, XM, YM, G, H, FI, DFI, KODE, NX)
C
C  SOLUTION OF THE SYSTEM OF EQUATIONS
C
      NN = 2*N
      CALL SLNPD(G, DFI, D, NN, NX)
C
C  COMPUTE STRESS AND DISPLACEMENT IN INTERNAL POINTS
C
      CALL INTER(FI, DFI, KODE, CX, CY, X, Y, SSOL, DSOL)
C
C  OUTPUT
C
      CALL OUTPT(XM, YM, FI, DFI, CX, CY, SSOL, DSOL)
      STOP
      END
```

Subroutine INPUT

This subroutine reads all the input required by the program. The data is presented in the following way,

(1) *Title card* Format 18A4, contains the name of the program.

(2) *Basic parameters card* (8I5, 2F10.4) Contains the number of elements, number of internal points, number of surfaces, last nodes of each different surface, the shear modulus and Poisson ratio.

(3) *Internal points coordinates cards* (2F10.4) As many cards as internal nodes are defined, each with the x_1x_2 coordinates of the node under consideration.

(4) *Extreme points of boundary elements cards* (2F10.4) Each card defines the coordinates of the extreme point of an element, read in the counterclockwise direction for external surfaces and clockwise for internal ones.

(5) *Boundary conditions cards* (I5, F10.4, I5, F10.4) As many cards as boundary nodes giving the values of the known variable in x_1 and x_2 directions. The variables are displacements if KODE = 0 or tractions if KODE = 1. KODE is read with I5 format and FI with F10.4.

The subroutine prints first the name of the job and the basic parameters. Then the coordinates of the extreme points of the elements and the boundary conditions given by node number, codes and prescribed values are output. The internal point coordinates will be printed in the subroutine OUTPT.

Program 19

```
      SUBROUTINE INPUT (CX, CY, X, Y, KODE, FI)
C
C PROGRAM 19
C
      COMMON N, L, NC(5), M, GE, XNU, LEC, IMP
      DIMENSION CX(1), CY(1), X(1), Y(1), KODE(1), FI(1), TITLE(18)
C
C N  =  NUMBER OF BOUNDARY ELEMENTS
C L  =  NUMBER OF INTERNAL POINTS WHERE THE FUNCTION IS
C       CALCULATED
C M  =  NUMBER OF DIFFERENT SURFACES
C NC(I) =  LAST NODE OF SURFACE I
C GE    =  SHEAR MODULUS
C XNU   =  POISSON MODULUS
C
      WRITE (IMP, 100)
  100 FORMAT (' ', 120('*'))
C
C  READ NAME OF THE JOB
C
      READ (LEC, 150) TITLE
  150 FORMAT (18A4)
      WRITE (IMP, 250) TITLE
  250 FORMAT (25X, 18A4)
C
C  READ BASIC PARAMETERS
C
      READ (LEC, 200) N, L, M, (NC(K), K = 1, 5), GE, XNU
  200 FORMAT (8I5, 2F10.4)
      WRITE (IMP, 300) N, L, GE, XNU
  300 FORMAT(//'DATA'//2X, 'NUMBER OF BOUNDARY ELEMENTS =',
     1 I3/2X, 'NUMBER OF INTERNAL POINTS WHERE THE FUNCTION IS',
     2 'CALCULATED =', I3/2X, 'SHEAR MODULUS =', E14.7/2X, 'POISSON',
     3 'MODULUS', E14.7)
      IF (M) 40, 40, 30
   30 WRITE (IMP, 999) M, (NC(K), K = 1, M)
  999 FORMAT (/2X, 'NUMBER OF DIFFERENT SURFACES =', I3/2X, 'LAST'
     1 'NODES IN THESE SURFACES: '/5(2X, I3))
C
C  READ INTERNAL POINTS COORDINATES
C
   40 DO 1 I = 1, L
    1 READ (LEC, 400) CX(I), CY(I)
  400 FORMAT (2F10.4)
C
C  READ COORDINATES OF EXTREME POINTS OF THE BOUNDARY
C  ELEMENTS IN ARRAY X AND Y
C
      WRITE (IMP, 500)
  500 FORMAT(//2X, 'COORDINATES OF THE EXTREME POINTS OF THE',
     1 'BOUNDARY ELEMENTS'//4X, 'POINT', 10X, 'X', 18X, 'Y')
      DO 10 I = 1, N
      READ (LEC, 600) X(I), Y(I)
  600 FORMAT (2F10.4)
   10 WRITE (IMP, 700) I, X(I), Y(I)
  700 FORMAT (5X, I3, 2(5X, E14.7))
```

```
C
C    READ BOUNDARY CONDITIONS
C    FI(I) = VALUE OF THE DISPLACEMENT IN THE COORDINATE I IF KODE = 0,
C    VALUE OF THE TRACTION IF KODE = 1.
C
        WRITE (IMP, 800)
  800   FORMAT(//2X, 'BOUNDARY CONDITIONS'//15X, 'PRESCRIBED',
       1 'VALUE', 15X, 'PRESCRIBED VALUE'/5X, 'NODE', 9X, 'X DIRECTION',
       2 8X, 'CODE', 8X, 'Y DIRECTION', 8X, 'CODE')
        DO 20 I = 1, N
        READ (LEC, 900) KODE(2*I-1), FI(2*I-1), KODE(2*I), FI(2*I)
  900   FORMAT (I5, F10.4, I5, F10.4)
   20   WRITE (IMP, 950)I, FI(2*I-1), KODE(2*I-1), FI(2*I), KODE(2*I)
  950   FORMAT (5X, I3, 8X, E14.7, 8X, I1, 8X, E14.7, 8X, I1)
        RETURN
        END
```

Subroutine FMAT

This subroutine computes the **G** and **H** matrices using the routines INTE and INLO.

> INTE: Computes the $\mathbf{H}_{ij}$, $\mathbf{G}_{ij}$ coefficients for the case $i \neq j$, using numerical integration.
> INLO: Calculates the matrix $\mathbf{G}_{ii}$.

The matrix $\mathbf{H}_{ii}$ is simply (smooth boundary)

$$\mathbf{H}_{ii} = \begin{bmatrix} \frac{1}{2} & 0 \\ 0 & \frac{1}{2} \end{bmatrix}$$

This subroutine also arranges the system of equations and prepares it to be solved. The matrix **A** of Equation (5.14) is stored now in **G**. Note that some rows in **A** have been multiplied by the shear modulus (GE) to avoid numerical error (i.e. all elements of matrix **G** are now of the same order). The right hand side vector **F** (stored in DFI) is also formed in this subroutine.

Program 20

```
      SUROUTINE FMAT(X, Y, XM, YM, G, H, FI, DFI, KODE, NX)
C
C   PROGRAM 20
C
C   THIS SUBROUTINE COMPUTES THE G AND H MATRICES AND FORM THE
C   SYSTEM A X = F
C
```

```
      COMMON N, L, NC(5), M, GE, XNU, LEC, IMP
      DIMENSION X(1), Y(1), XM(1), YM(1), G(NX, NX), H(NX, NX), FI(1)
      DIMENSION KODE(1), DFI(1)
C
C  COMPUTE THE MID-POINT COORDINATES AND STORE IN ARRAY XM
C  AND YM
C
      X(N + 1) = X(1)
      Y(N + 1) = Y(1)
      DO 10 I = 1, N
      XM(I) = (X(I) + X(I + 1))/2
   10 YM(I) = (Y(I) + Y(I + 1))/2
      IF(M - 1)15, 15, 12
   12 XM(NC(1)) = (X(NC(1)) + X(1))/2
      YM(NC(1)) = (Y(NC(1)) + Y(1))/2
      DO 13 K = 2, M
      XM(NC(K)) = (X(NC(K)) + X(NC(K - 1) + 1))/2
   13 YM(NC(K)) = (Y(NC(K)) + Y(NC(K - 1) + 1))/2
C
C  COMPUTE G AND H MATRICES
C
   15 DO 30 I = 1, N
      DO 30 J = 1, N
      IF(M - 1)16, 16, 17
   17 IF(J - NC(1))19, 18, 19
   18 KK = 1
      GO TO 23
   19 DO 22 K = 2, M
      IF(J - NC(K))22, 21, 22
   21 KK = NC(K - 1) + 1
      GO TO 23
   22 CONTINUE
   16 KK = J + 1
   23 IF(I - J)20, 25, 20
   20 CALL INTE(XM(I), YM(I), X(J), Y(J), X(KK), Y(KK), H((2*I - 1), (2*J - 1))
     1 H((2*I - 1), (2*J)), H((2*I), (2*J - 1)), H((2*I), (2*J)), G((2*I - 1),
     2 (2*J - 1)), G((2*I - 1), (2*J)), G((2*I), (2*J)))
      G((2*I), (2*J - 1)) = G((2*I - 1), (2*J))
      GO TO 30
   25 CALL INLO(X(J), Y(J), X(KK), Y(KK), G((2*I - 1), (2*J - 1)), G((2*I - 1),
     1 (2*J)), G((2*I), (2*J)))
      H((2*I - 1), (2*J - 1)) = 0.5
      H((2*I), (2*J)) = 0.5
      H((2*I - 1), (2*J)) = 0.
      H((2*I), (2*J - 1)) = 0.
      G((2*I), (2*J - 1)) = G((2*I - 1), (2*J))
   30 CONTINUE
C
C  ARRANGE THE SYSTEM OF EQUATIONS READY TO BE SOLVED
C
      NN = 2*N
      DO 50 J = 1, NN
      IF(KODE(J))43, 43, 40
   40 DO 42 I = 1, NN
      CH = G(I, J)
      G(I, J) = -H(I, J)
```

```
   42 H(I, J) = -CH
      GO TO 50
   43 DO 45 I = 1, NN
   45 G(I, J) = G(I, J)*GE
   50 CONTINUE
C
C  DFI ORIGINALLY CONTAINS THE INDEPENDENT COEFFICIENTS,
C  AFTER SOLUTION IT WILL CONTAIN THE VALUES OF THE SYSTEM
C  UNKNOWNS
C
      DO 60 I = 1, NN
      DFI(I) = 0.
      DO 60 J = 1, NN
      DFI(I) = DFI(I) + H(I, J)*FI(J)
   60 CONTINUE
      RETURN
      END
```

Subroutine INTE

Computes the values of **H** and **G** coefficients using numerical integration. Note that $H_{12} \neq H_{21}$ but $G_{12} = G_{21}$.

Program 21

```
      SUBROUTINE INTE(XP, YP, X1, Y1, X2, Y2, H11, H12, H21, H22, G11,
     1 G12, G22)
C
C  PROGRAM 21
C
C  THIS SUBROUTINE COMPUTES THE VALUES OF THE H AND G MATRIX
C  ELEMENTS THAT RELATE A NODE WITH A DIFFERENT ONE, BY MEANS
C  OF NUMERICAL INTEGRATION ALONG THE BOUNDARY ELEMENTS
C
C
C  DIST = DISTANCE FROM THE POINT UNDER CONSIDERATION TO THE
C  BOUNDARY ELEMENTS
C  RA = DISTANCE FROM THE POINT UNDER CONSIDERATION TO THE
C  INTEGRATION POINTS IN THE BOUNDARY ELEMENTS
C
      COMMON N, L, NC(5), M, GE, XNU, LEC, IMP
      DIMENSION XCO(4), YCO(4), GI(4), OME(4)
      GI(1) = 0.86113631
      GI(2) = -GI(1)
      GI(3) = 0.33998104
      GI(4) = -GI(3)
      OME(1) = 0.34785484
      OME(2) = OME(1)
      OME(3) = 0.65214515
      OME(4) = OME(3)
      AX = (X2 - X1)/2
      BX = (X2 + X1)/2
```

```
      AY = (Y2 - Y1)/2
      BY = (Y2 + Y1)/2
      ETA1 = (Y2 - Y1)/(2*SQRT(AX**2 + AY**2))
      ETA2 = (X1 - X2)/(2*SQRT(AX**2 + AY**2))
      IF(AX)10, 20, 10
   10 TA = AY/AX
      DIST = ABS((TA*XP - YP + Y1 - TA*X1)/SQRT(TA**2 + 1))
      GO TO 30
   20 DIST = ABS(XP - X1)
   30 SIG = (X1 - XP)*(Y2 - YP) - (X2 - XP)*(Y1 - YP)
      IF(SIG)31, 32, 32
   31 DIST = -DIST
   32 H11 = 0.
      H12 = 0.
      H21 = 0.
      H22 = 0.
      G11 = 0.
      G12 = 0.
      G22 = 0.
      DE = 4*3.141592*(1 - XNU)
      DO 40 I = 1, 4
      XCO(I) = AX*GI(I) + BX
      YCO(I) = AY*GI(I) + BY
      RA = SQRT((XP - XCO(I))**2 + (YP - YCO(I))**2)
      RD1 = (XCO(I) - XP)/RA
      RD2 = (YCO(I) - YP)/RA
      G11 = G11 + ((3 - 4* XNU)*ALOG(1/RA) + RD1**2)*OME(I)*SQRT
     1 (AX**2 + AY**2)/(2*DE*GE)
      G12 = G12 + RD1*RD2*OME(I)*SQRT(AX**2 + AY**2)/(2*DE*GE)
      G22 = G22 + ((3 - 4*XNU)*ALOG(1/RA) + RD2**2)*OME(I)*SQRT
     1 (AX**2 + AY**2)/(2*DE*GE)
      H11 = H11 - DIST*((1 - 2*XNU) + 2*RD1**2)/(RA**2*DE)*OME(I)
     1 *SQRT(AX**2 + AY**2)
      H12 = H12 - (DIST*2*RD1*RD2/RA + (1 - 2*XNU)*(ETA1*RD2 -
     1 ETA2*RD1))*OME(I)*SQRT(AX**2 + AY**2)/(RA*DE)
      H21 = H21 - (DIST*2*RD1*RD2/RA + (1 - 2*XNU)*(ETA2*RD1 -
     1 ETA1*RD2))*OME(I)*SQRT(AX**2 + AY**2)/(RA*DE)
   40 H22 = H22 - DIST*((1 - 2*XNU) + 2*RD2**2)*OME(I)*SQRT(AX**2 +
     1 AY**2)/(RA**2*DE)
      RETURN
      END
```

Subroutine INLO

Calculates the $\mathbf{G}_{ij}$ matrix. The integration here has been done analytically but it could also have been calculated using numerical integration.

Program 22

```
      SUBROUTINE INLO(X1, Y1, X2, Y2, G11, G12, G22)
C
C PROGRAM 22
C
```

```
C  THIS SUBROUTINE COMPUTES THE VALUES OF THE ELEMENTS OF THE G
C  MATRIX THAT RELATE AN ELEMENT WITH ITSELF
C
      COMMON N, L, NC(5), M, GE, XNU, LEC, IMP
      AX = (X2 - X1)/2
      AY = (Y2 - Y1)/2
      SR = SQRT(AX**2 + AY**2)
      DE = 4*3.141492*GE*(1 - XNU)
      G11 = SR*((3 - 4*XNU)*(1 - ALOG(SR)) + (X2 - X1)**2/(4*SR**2))/DE
      G22 = SR*( (3 - 4*XNU)*(1 - ALOG(SR)) + (Y2 - Y1)**2/(4*SR**2))/DE
      G12 = (X2 - X1)*(Y2 - Y1)/(4*SR*DE)
      RETURN
      END
```

Subroutine INTER

This subroutine first reorders the vectors DFI and FI in such a way that all the boundary displacements are stored in FI and all tractions in DFI. It then computes the stresses and displacements at internal points. In order to do so, it uses formulae (5.15) and (5.16). The integrals of the coefficients *S* and *D* given by the last formula are computed in another routine called SIGM.

Program 23

```
      SUBROUTINE INTER(FI, DFI, KODE, CX, CY, X, Y, SSOL, DSOL)
C
C  PROGRAM 23
C
C  THIS SUBROUTINE COMPUTES THE STRESSES AND DISPLACEMENTS AT
C  INTERNAL POINTS.
C
      COMMON N, L, NC(5), M, GE, XNU, LEC, IMP
      DIMENSION FI(1), DFI(1), KODE(1), CX(1), CY(1), X(1), Y(1), SSOL(1)
      DIMENSION DSOL(1)
C
C  REORDER FI AND DFI ARRAY TO PUT ALL THE DISPLACEMENTS IN
C  FI AND ALL THE TRACTIONS IN DFI
C
      NN = 2*N
      DO 20 I = 1, NN
      IF(KODE(I)) 15, 15, 10
   10 CH = FI(I)
      FI(I) = DFI(I)
      DFI(I) = CH
      GO TO 20
   15 DFI(I) = DFI(I)*GE
   20 CONTINUE
C
C  COMPUTE STRESS AND DISPLACEMENT IN INTERNAL POINTS
C
```

```
      DO 40 K = 1, L
      DSOL(2*K - 1) = 0.
      DSOL(2*K) = 0.
      SSOL(3*K - 2) = 0.
      SSOL(3*K - 1) = 0.
      SSOL(3*K) = 0.
      DO 30 J = 1, N
      IF(M - 1)28, 28, 22
   22 IF(J - NC(1))24, 23, 24
   23 KK = 1
      GO TO 29
   24 DO 26 LK = 2, M
      IF(J - NC(LK))26, 25, 26
   25 KK = NC(LK - 1) + 1
      GO TO 29
   26 CONTINUE
   28 KK = J + 1
   29 CALL INTE(CX(K), CY(K), X(J), Y(J), X(KK), Y(KK), H11, H12, H21, H22,
     1 G11, G12, G22)
      DSOL(2*K - 1) = DSOL(2*K - 1) + DFI(2*J - 1)*G11 + DFI(2*J)*G12 - FI
     1 (2*J - 1)*H11 - FI(2*J)*H12
      DSOL(2*K) = DSOL(2*K) + DFI(2*J - 1)*G12 + DFI(2*J)*G22 - FI
     1 (2*J - 1)*H21 - FI(2*J)*H22
      CALL SIGM(CX(K), CY(K), X(J), Y(J), X(KK), Y(KK), D111, D211, D112,
     1 D212, D122, D222, S111, S211, S112, S212, S122, S222)
      SSOL(3*K - 2) = SSOL(3*K - 2) + DFI(2*J - 1)*D111 + DFI(2*J)*D211 -
     1 FI(2*J - 1)*S111 - FI(2*J)*S211
      SSOL(3*K - 1) = SSOL(3*K - 1) + DFI(2*J - 1)*D112 + DFI(2*J)*D212 -
     1 FI(2*J - 1)*S112 - FI(2*J)*S212
   30 SSOL(3*K) = SSOL(3*K) + DFI(2*J - 1)*D122 + DFI(2*J)*D222 - FI
     1 (2*J - 1)*S122 - FI(2*J)*S222
   40 CONTINUE
      RETURN
      END
```

Subroutine SIGM

The D and S coefficients needed to calculate the internal stresses are given according to formulae (4.82) and (4.83) and the integrals of the coefficients are evaluated using Gaussian quadrature.

Program 24

```
      SUBROUTINE SIGM(XP, YP, X1, Y1, X2, Y2, D111, D211, D112, D212,
     1 D122, D222, S111, S211, S112, S212, S122, S222)
C
C  PROGRAM 24
C
C  THIS SUBROUTINE COMPUTES THE VALUES OF THE S AND D MATRICES
C  IN ORDER TO COMPUTE THE STRESS IN ANY INTERNAL POINT
C
```

```
      COMMON N, L, NC(5), M, GE, XNU, LEC, IMP
      DIMENSION XCO(4), YCO(4), GI(4), OME(4)
      GI(1) = 0.86113631
      GI(2) = -GI(1)
      GI(3) = 0.33998104
      GI(4) = -GI(3)
      OME(1) = 0.34785485
      OME(2) = OME(1)
      OME(3) = 0.65214515
      OME(4) = OME(3)
      AX = (X2 - X1)/2
      BX = (X2 + X1)/2
      AY = (Y2 - Y1)/2
      BY = (Y2 + Y1)/2
      ETA1 = (Y2 - Y1)/(2*SQRT(AX**2 + AY**2))
      ETA2 = (X1 - X2)/(2*SQRT(AX**2 + AY**2))
      IF(AX) 10, 20, 10
10    TA = AY/AX
      DIST = ABS((TA*XP - YP + Y1 - TA*X1)/SQRT(TA**2 + 1))
      GO TO 30
20    DIST = ABS(XP - X1)
30    SIG = (X1 - XP)*(Y2 - YP) - (X2 - XP)*(Y1 - YP)
      IF(SIG)31, 32, 32
31    DIST = -DIST
32    D111 = 0.
      D211 = 0.
      D112 = 0.
      D212 = 0.
      D122 = 0.
      D222 = 0.
      S111 = 0.
      S211 = 0
      S112 = 0.
      S212 = 0.
      S122 = 0.
      S222 = 0.
      FA = 1 - 4*XNU
      AL = 1 - 2*XNU
      DE = 4*3.141592*(1 - XNU)
      DO 40 I = 1, 4
      XCO(I) = AX*GI(I) + BX
      YCO(I) = AY*GI(I) + BY
      RA = SQRT((XP - XCO(I))**2 + (YP - YCO(I))**2)
      RD1 = (XCO(I) - XP)/RA
      RD2 = (YCO(I) - YP)/RA
      D111 = D111 + (AL*RD1 + 2*RD1**3)*OME(I)*SQRT(AX**2 + AY**2)/
     1 (DE*RA)
      D211 = D211 + (2*RD1**2*RD2 - AL*RD2)*OME(I)*SQRT(AX**2 +
     1 AY**2)/(DE*RA)
      D112 = D112 + (AL*RD2 + 2*RD1**2*RD2)/(DE*RA)*OME(I)*SQRT
     1 (AX**2 + AY**2)
      D212 = D212 + (AL*RD1 + 2*RD1 *RD2**2)/(DE*RA)*OME(I)*SQRT
     1 (AX**2 + AY**2)
      D122 = D122 + (2*RD1*RD2**2 - AL*RD1)/(DE*RA)*OME(I)*SQRT
     1 (AX**2 + AY**2)
```

```
      D222 = D222 + (AL*RD2 + 2*RD2**3)/(DE*RA)*OME(I)*SQRT(AX**2 +
     1 AY**2)
      S111 = S111 + (2*DIST/RA*(AL*RD1 + XNU*2*RD1 - 4*RD1**3) +
     1 4*XNU*ETA1*RD1**2 + AL*(2*ETA1*RD1**2 + 2*ETA1) - FA*ETA1)
     2 *2*GE/(DE*RA**2)*OME(I)*SQRT(AX**2 + AY**2)
      S211 = S211 + (2*DIST/RA*(AL*RD2 - 4*RD1**2*RD2) + 4*XNU*ETA1
     1 *RD1*RD2 + AL*2*ETA2*RD1**2 - FA*ETA2)*2*GE/(DE*RA**2)
     2 *OME(I)*SQRT(AX**2 + AY**2)
      S112 = S112 + (2*DIST/RA*(XNU*RD2 - 4*RD1**2*RD2) + 2*XNU*
     1 (ETA1*RD2*RD1 + ETA2*RD1**2) + AL*(2*ETA1*RD1*RD2 + ETA2))*2*G
     2 (DE*RA**2)*OME(I)*SQRT(AX**2 + AY**2)
      S212 = S212 + (2*DIST/RA*(XNU*RD1 - 4*RD1*RD2**2) + 2*XNU*
     1 (ETA1*RD2**2 + ETA2*RD1*RD2) + AL*(2*ETA2*RD1*RD2 + ETA1))
     2 *2*GE/(DE*RA**2)*OME(I)*SQRT(AX**2 + AY**2)
      S122 = S122 + (2*DIST/RA*(AL*RD1 - 4*RD1*RD2**2) + 4*XNU*
     1 ETA2*RD1*RD2 + AL*2*ETA1*RD2**2 - FA*ETA1)*2*GE/(DE*RA**2)
     2 *OME(I)*SQRT(AX**2 + AY**2)
   40 S222 = S222 + (2*DIST/RA*(AL*RD2 + 2*XNU*RD2 - 4*RD2* *3) +
     1 4*XNU*ETA2*RD2**2 + AL*(2*ETA2*RD2**2 + 2*ETA2) - FA*ETA2)
     2 *2*GE/(DE*RA**2) *OME(I)*SQRT(AX**2 + AY**2)
      RETURN
      END
```

Subroutine OUTPUT

Results are printed in the following order.

(1) *Boundary nodes* with coordinates x_1x_2, values of u_1u_2 displacements and tractions p_1p_2.

(2) *Internal nodes* with coordinates x_1x_2, values of u_1u_2 displacements and stresses σ_{11}, σ_{12}, σ_{22}.

Program 25

```
      SUBROUTINE OUTPT(XM, YM, FI, DFI, CX, CY, SSOL, DSOL)
C
C PROGRAM 25
C
      COMMON N, L, NC(5), M, GE, XNU, LEC, IMP
      DIMENSION XM(1), YM(1), FI(1), DFI(1), CX(1), CY(1), SSOL(1), DSOL(1)
      WRITE(IMP, 100)
  100 FORMAT(' ', 120('*')//1X, 'RESULTS'//2X, 'BOUNDARY NODES'//11X,
     1 'X', 18X, 'Y', 12X, 'DISPLACEMENT X', 5X, 'DISPLACEMENT Y', 7X,
     2 'TRACTION X', 9X, 'TRACTION Y'/)
      DO 10 I = 1, N
   10 WRITE(IMP, 200) XM(I), YM(I), FI(2*I - 1), FI(2*I), DFI(2*I - 1),
     1 DFI(2*I)
  200 FORMAT(6(5X, E14.7))
      WRITE(IMP, 300)
```

```
300   FORMAT(//2X, 'INTERNAL POINTS'//9X, 'X', 16X, 'Y', 10X, 'DISPLACE-',
     1 'MENT X', 3X, 'DISPLACEMENT Y', 6X, 'SIGMA X', 11X, 'TAU XY', 10X,
     2 'SIGMA Y')
      DO 20 K = 1, L
 20   WRITE(IMP, 400) CX(K), CY(K), DSOL(2*K - 1), DSOL(2*K), SSOL(3*K - 2),
     1 SSOL(3*K - 1), SSOL(3*K)
400   FORMAT(7(3X, E14.7))
      WRITE(IMP, 500)
500   FORMAT(' ', 120('*'))
      RETURN
      END
```

Example 5.1

The following example shows the application of boundary elements for the case of a circular cavity under internal pressure in an infinite medium.

The boundary has been divided into 24 constant elements and equal number of nodes and there are 10 internal nodes (see Figure 5.4), where displacements and stresses are computed.

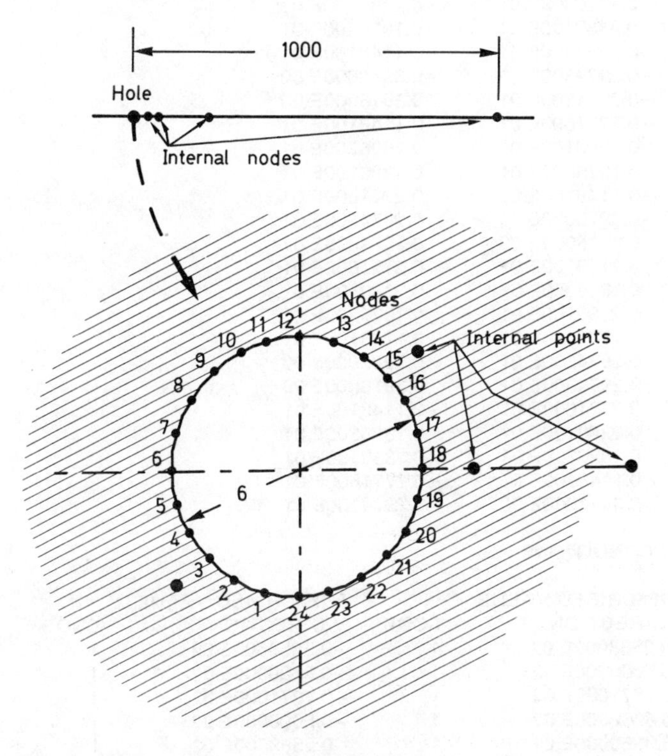

Fig. 5.4. Circular cavity under internal pressure

To avoid rigid body motion a series of displacement components were prescribed equal to zero. They are the displacements in the x_1 direction at nodes 12 and 24 and the displacements in the x_2 direction at node 18.

The symmetry of the results is clearly seen in the following listing.

Test of the CEBE program, circular hole

DATA

NUMBER OF BOUNDARY ELEMENTS = 24
NUMBER OF INTERNAL POINTS WHERE THE FUNCTION IS CALCULATED = 10
SHEAR MODULUS = 0.9450000E 05
POISSON MODULUS 0.1000000E 00

COORDINATES OF THE EXTREME POINTS OF THE BOUNDARY ELEMENTS

POINT	X	Y
1	−0.3916000E 00	−0.2974300E 01
2	−0.1148100E 01	−0.2771600E 01
3	−0.1826300E 01	−0.2380100E 01
4	−0.2380100E 01	−0.1825300E 01
5	−0.2771600E 01	−0.1148100E 01
6	−0.2974300E 01	−0.3916000E 00
7	−0.2974300E 01	0.3916000E 00
8	−0.2771600E 01	0.1148100E 01
9	−0.2380100E 01	0.1826300E 01
10	−0.1826300E 01	0.2380100E 01
11	−0.1148100E 01	0.2771600E 01
12	−0.3916000E 00	0.2974300E 01
13	0.3916000E 00	0.2974300E 01
14	0.1148100E 01	0.2771600E 01
15	0.1826300E 01	0.2380100E 01
16	0.2380100E 01	0.1826300E 01
17	0.2771600E 01	0.1148100E 01
18	0.2974300E 01	0.3916000E 00
19	0.2974300E 01	−0.3916000E 00
20	0.2771600E 01	−0.1148100E 01
21	0.2380100E 01	−0.1826300E 01
22	0.1826300E 01	−0.2380100E 01
23	0.1148100E 01	−0.2771600E 01
24	0.3916000E 00	−0.2974300E 01

BOUNDARY CONDITIONS

NODE	PRESCRIBED VALUE X DIRECTION	CODE	PRESCRIBED VALUE Y DIRECTION	CODE
1	−0.2588000E 02	1	−0.9659000E 02	1
2	−0.5000000E 02	1	−0.8660000E 02	1
3	−0.7071000E 02	1	−0.7071000E 02	1
4	−0.8660000E 02	1	−0.5000000E 02	1
5	−0.9659000E 02	1	−0.2588000E 02	1
6	−0.1000000E 03	1	0.0000000E 00	1

7	−0.9659000E 02	1	0.2588000E 02	1
8	−0.8660000E 02	1	0.5000000E 02	1
9	−0.7071000E 02	1	0.7071000E 02	1
10	−0.5000000E 02	1	0.8660000E 02	1
11	−0.2588000E 02	1	0.9659000E 02	1
12	0.0000000E 00	0	0.1000000E 03	1
13	0.2588000E 02	1	0.9659000E 02	1
14	0.5000000E 02	1	0.8660000E 02	1
15	0.7071000E 02	1	0.7071000E 02	1
16	0.8660000E 02	1	0.5000000E 02	1
17	0.9659000E 02	1	0.2588000E 02	1
18	0.1000000E 03	1	0.0000000E 00	0
19	0.9659000E 02	1	−0.2588000E 02	1
20	0.8660000E 02	1	−0.5000000E 02	1
21	0.7071000E 02	1	−0.7071000E 02	1
22	0.5000000E 02	1	−0.8660000E 02	1
23	0.2588000E 02	1	−0.9659000E 02	1
24	0.0000000E 00	0	−0.1000000E 03	1

RESULTS

BOUNDARY NODES

X	Y	DISPLACEMENT X	DISPLACEMENT Y	TRACTION X	TRACTION Y
−0.7698500E 00	−0.2872950E 01	−0.4244857E − 03	−0.1584250E − 02	−0.2588000E 02	−0.9659000E 02
−0.1487200E 01	−0.2575750E 01	−0.8200233E − 03	−0.1420372E − 02	−0.5000000E 02	−0.8660000E 02
−0.2103200E 01	−0.2103200E 01	−0.1159720E − 02	−0.1159720E − 02	−0.7071000E 02	−0.7071000E 02
−0.2575850E 01	−0.1487200E 01	−0.1420372E − 02	−0.8200233E − 03	−0.8660000E 02	−0.5000000E 02
−0.2872950E 01	−0.7698500E 00	−0.1584250E − 02	−0.4244857E − 03	−0.9659000E 02	−0.2588000E 02
−0.2974300E 01	0.0000000E 00	−0.1640153E − 02	−0.7105427E − 14	−0.1000000E 03	0.0000000E 00
−0.2872950E 01	0.7698500E 00	−0.1584250E − 02	0.4244857E − 03	−0.9659000E 02	0.2588000E 02
−0.2575850E 01	0.1487200E 01	−0.1420372E − 02	0.8200233E − 03	−0.8660000E 02	0.5000000E 02
−0.2103200E 01	0.2103200E 01	−0.1159720E − 02	0.1159720E − 02	−0.7071000E 02	0.7071000E 02
−0.1487200E 01	0.2575850E 01	−0.8200233E − 03	0.1420372E − 02	−0.5000000E 02	0.8660000E 02
−0.7698500E 00	0.2872850E 01	−0.4244857E − 03	0.1584250E − 02	−0.2588000E 02	0.9659000E 02
0.0000000E 00	0.2974300E 01	0.0000000E 00	0.1640153E − 02	0.4448442E − 08	0.1000000E 03
0.7698500E 00	0.2872950E 01	0.4244857E − 03	0.1584250E − 02	0.2588000E 02	0.9659000E 02
0.1487200E 01	0.2575850E 01	0.8200233E − 03	0.1420372E − 02	0.5000000E 02	0.8660000E 02
0.2130200E 01	0.2103200E 01	0.1159720E − 02	0.1159720E − 02	0.7071000E 02	0.7071000E 02
0.2575850E 01	0.1487200E 01	0.1420372E − 02	0.8200233E − 03	0.8660000E 02	0.5000000E 02
0.2872950E 01	0.7698500E 00	0.1584250E − 02	0.4244857E − 03	0.9659000E 02	0.2588000E 02
0.2974500E 01	0.0000000E 00	0.1640153E − 02	0.0000000E 00	0.1000000E 02	0.2266187E − 08
0.2872950E 01	−0.7698500E 00	0.1584250E − 02	−0.4244857E − 03	0.9659857E 02	−0.2588000E 02
0.2575850E 01	−0.1487200E 01	0.1420372E − 02	−0.8200233E − 03	0.8660000E 02	−0.5000000E 02
0.2103200E 01	−0.2103200E 01	0.1159720E − 02	−0.1159720E − 02	0.7071000E 02	−0.7071000E 02
0.1487200E 01	−0.2575850E 01	0.8200233E − 03	−0.1420372E − 02	0.5000000E 02	−0.8660000E 02
0.7698500E 00	−0.2872950E 01	0.4244857E − 03	−0.1584250E − 02	0.2588000E 02	−0.9659000E 02
0.0000000E 00	−0.2974300E 01	0.0000000E 00	−0.1640153E − 02	0.9442447E − 09	−0.1000000E 03

INTERNAL POINTS

X	Y	DISPLACEMENT X	DISPLACEMENT Y	SIGMA X	TAU XY	SIGMA Y
0.4000000E 01	0.0000000E 00	0.1204831E − 02	0.2198242E − 13	−0.5723506E 02	−0.1538865E − 08	0.5711798E 02
0.2828430E 01	0.2828430E 01	0.8519233£ − 03	0.8519233E − 03	−0.5609629E − 01	−0.5717716E 02	−0.5609629E − 01
−0.4000000E 01	0.0000000E 00	−0.1204831E − 02	0.9325873E − 14	−0.5723506E 02	−0.8694769E − 09	0.5711798E 02
−0.2828430E 01	−0 2828450E 01	−0.8519233E − 03	−0.8519233E − 03	−0.5609629E − 01	−0.5717716E 02	−0.5609629E − 01
0.6000000E 01	0.0000000E 00	0.8030049E − 03	0.4773959E − 14	−0.2529407E 02	−0.2564775E − 09	0.2529451E 02
0.1000000E 02	0.0000000E 00	0.4818006E − 03	0.9769963E − 14	−0.9106095E 01	0.5184120E − 10	0.9106025E 01
0.2000000E 02	0.0000000E 00	0.2408997E − 03	−0.4794776E − 14	−0.2276507E 01	0.3308287E − 10	0.2276502E 01
0.5000000E 02	0.0000000E 00	0.9635982E − 04	0.2260545E − 13	−0.3642402E 00	−0.4959588E − 11	0.3642401E 00
0.2000000E 03	0.0000000E 00	0.2408995E − 04	−0.2949138E − 14	−0.2276500E − 01	0.3763656E − 12	0.2276500E − 01
0.1000000E 04	0.0000000E 00	0.4817990E − 05	−0.4250885E − 13	−0.9106001E − 03	0.1464384E − 12	0.9106001E − 03

Table 5.1. Radial stresses at internal points (Example 5.1)

Distance to the centre of the cavity	*Boundary element method*	*Elasticity theory*
4	−57.23	−56.25
6	−25.29	−25.00
10	−9.10	−9.00
20	−2.27	−2.25
50	−0.36	−0.36
200	-0.227×10^{-1}	-0.225×10^{-1}
1000	-0.911×10^{-3}	-0.9×10^{-3}

As was expected displacements and stresses decay with increasing distance from the cavity. The radial and hoop stresses at these internal points have the same absolute value but different sign which is correct. Theoretical results are compared against the boundary element solution in Table 5.1 and the agreement is satisfactory.

Example 5.2

The following application compares the boundary elements against finite elements results and the exact elasticity solution for a 90° segment of a pipe under internal pressure (Figure 5.5).

The finite element mesh shown in Figure 5.5(a) consists of 52 nodes and 76 three-noded elements, while the boundary element discretization has 26 constant segments (Figure 5.5(b)). Results were obtained for displacements and circumferential and radial stresses. The finite element stresses were computed at the centre of each element. Results for stresses are compared in Figure 5.6 where the boundary element solution agrees very well with the exact solution. The agreement between finite elements and the exact results is instead very poor and illustrates the dangers of using this type of element (constant strain) in elasticity. It should be pointed out that boundary element stresses very close to the boundary are also unsatisfactory as is to be expected. As a rule errors occur for all internal points that are situated at a distance from the boundary less than half an element length.

Although the disagreement for stresses between both methods is considerable, the displacements agree reasonably well. For instance, at finite element internal nodes 26 and 27 the results (Table 5.2) for the displacement in the x_1 or x_2 direction (note that $u_1 = u_2$ at 45°) are similar for both methods.

Example 5.3

The following example shows how the boundary element method can be applied to study the stress field surrounding an anchor plate. The plate is

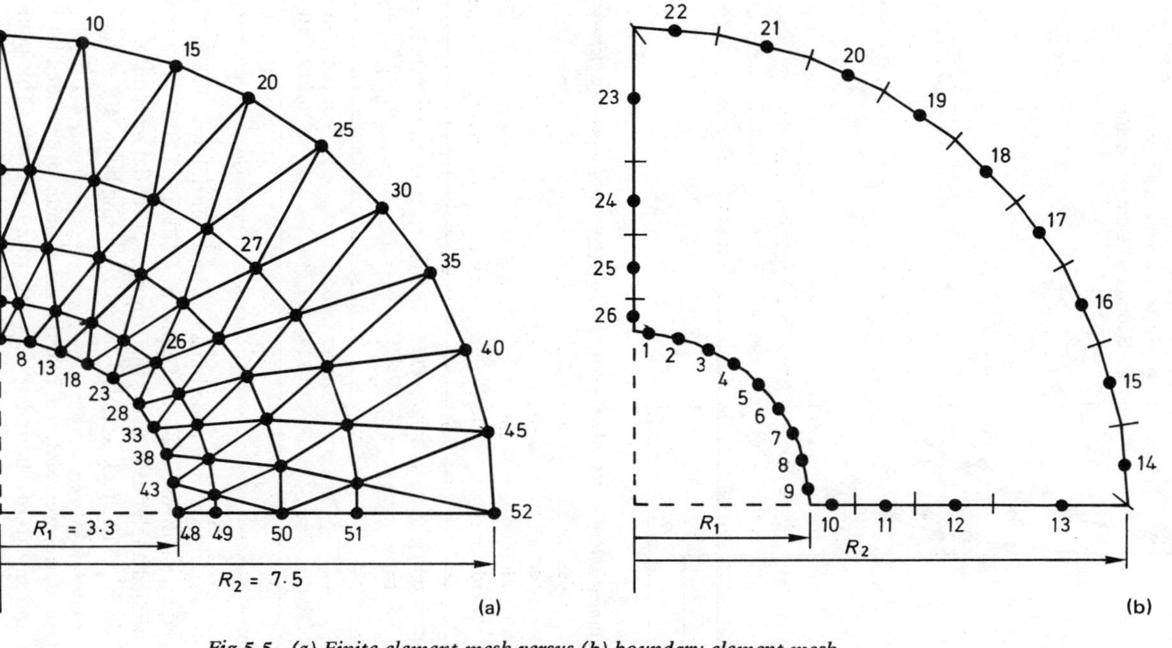

Fig. 5.5. (a) Finite element mesh versus (b) boundary element mesh

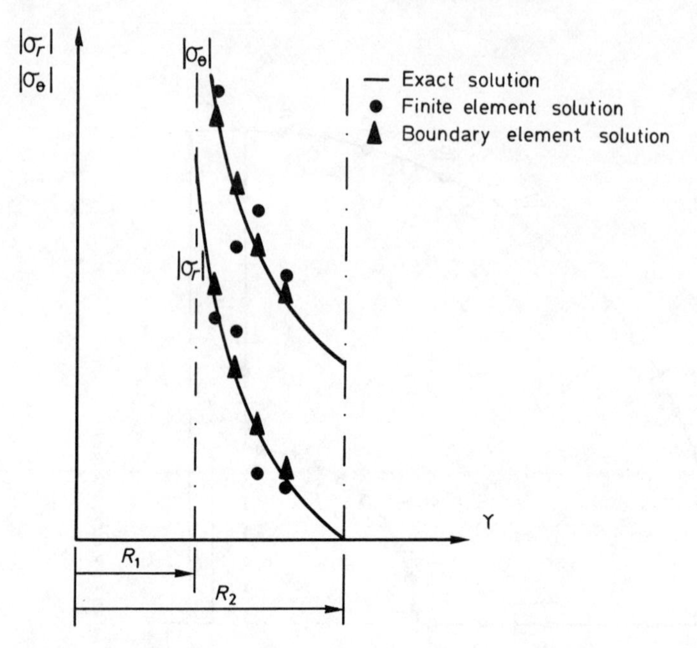

Fig. 5.6. Comparison of exact solution with finite element and boundary element solution

Table 5.2

Node	*Finite element result*	*Boundary element result*
26	0.157×10^{-3}	0.164×10^{-3}
27	0.124×10^{-3}	0.129×10^{-3}

assumed to be rigid and embedded in an infinite continuum. The loading is produced by an anchor cable and equal to P (Figure 5.7(a)). The problem was studied analytically using the complex potential method of Muskheliskvili[1]. The crack was also discretized using boundary elements as shown in Figure 5.7(b).

Two boundaries were defined, one the crack boundary divided into 74 elements and an external boundary very far from the crack. This boundary is at radius of 50 crack lengths and was needed in order to restrain rigid body modes.

The results obtained using boundary elements can be seen in Figures 5.8(a) and 5.8(b). In the first figure the tractions over the lower surface of

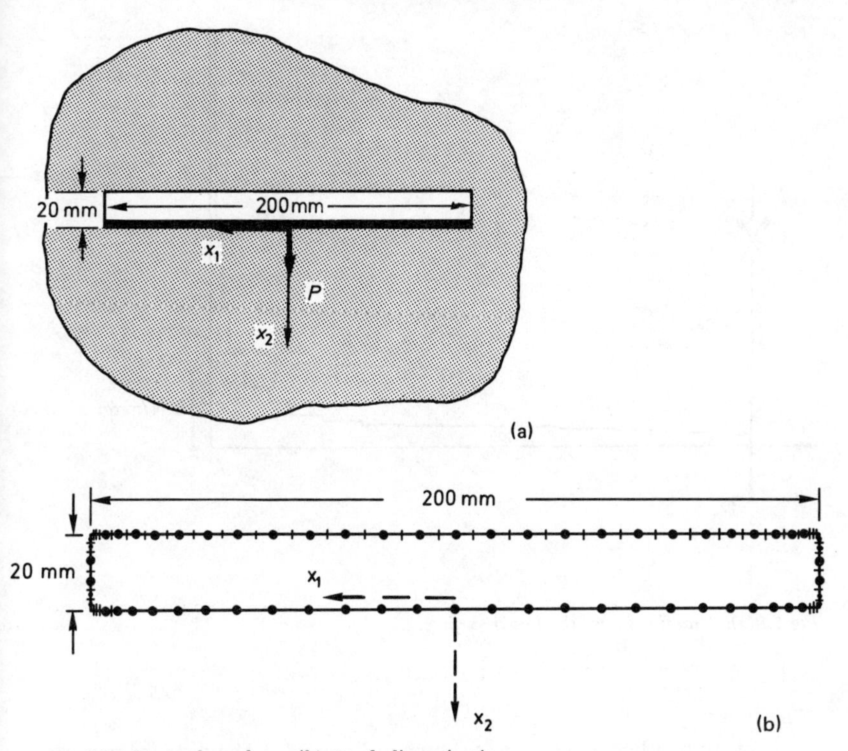

Fig. 5.7. (a) Anchor plate. (b) Crack discretization

the crack (i.e. where the rigid plate is acting) are plotted and their shape and values agree well with the analytical results. Figure 5.8(b) shows the internal stresses σ_{11} and σ_{22} along the axes x_2 (i.e. $x_1 = 0$). As a comparison the results for an internal point situated at $x_1 = 0$ and $x_2 = 30$ mm are presented in Table 5.3. They correspond to a total force of 1000 N in the anchor cable. The agreement is very satisfactory and it points out the convenience of using boundary elements for stress concentration problems such as this.

5.3 LINEAR AND HIGHER ORDER ELEMENTS

Consider the basic boundary equation at a point 'i', without body forces for simplicity, i.e.

$$cu^i + \int_\Gamma p^* u \, d\Gamma = \int_\Gamma u^* p \, d\Gamma \tag{5.20}$$

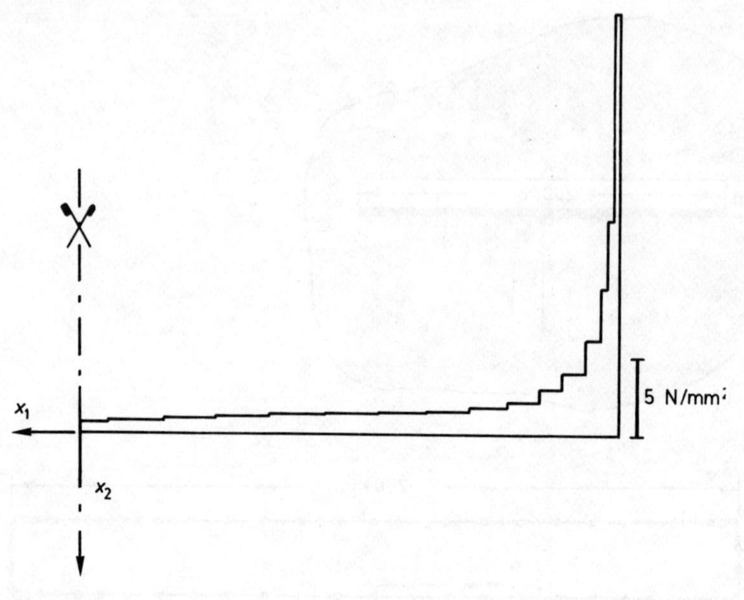

Fig.5.8(a). Tractions over the lower surface.

If we study a linear element as shown in Figure 5.9, the displacements and tractions can be written as,

$$\mathbf{u} = \begin{Bmatrix} u_1 \\ u_2 \end{Bmatrix} = \begin{bmatrix} \phi_1 & \phi_2 & 0 & 0 \\ 0 & 0 & \phi_1 & \phi_2 \end{bmatrix} \begin{Bmatrix} u_1^1 \\ u_1^2 \\ u_2^1 \\ u_2^2 \end{Bmatrix} = \mathbf{\Phi}^T \mathbf{u}_j \tag{5.21}$$

$$\mathbf{p} = \begin{Bmatrix} p_1 \\ p_2 \end{Bmatrix} = \begin{bmatrix} \phi_1 & \phi_2 & 0 & 0 \\ 0 & 0 & \phi_1 & \phi_2 \end{bmatrix} \begin{Bmatrix} p_1^1 \\ p_1^2 \\ p_2^1 \\ p_2^2 \end{Bmatrix} = \mathbf{\Phi}^T \mathbf{p}_j \tag{5.22}$$

(u_j, p_j refers to the nodal displacement and traction on the element j.)

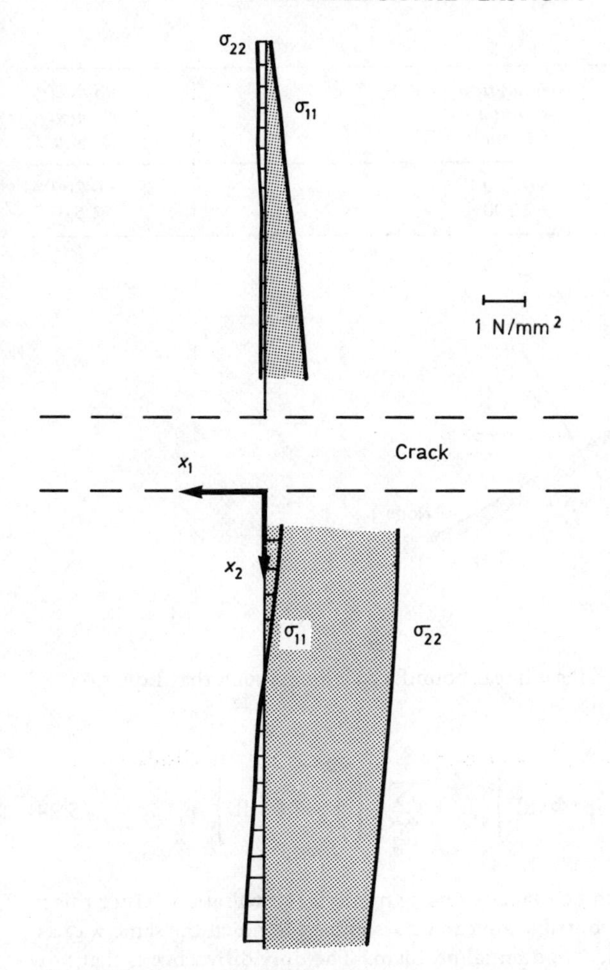

Fig. 5.8(b). Stresses along the x_2 axis.

The function ϕ_1 are linear interpolation functions, such that

$$\phi_1 = -\frac{1}{2}(\xi - 1)$$

$$\phi_2 = \frac{1}{2}(\xi + 1) \qquad (5.23)$$

Note that the functions for u and p are both linear. This need not always be the case, and it is more consistent to take the functions for u one degree higher than those for p.

Table 5.3

	Analytical solution (N/mm²)	*Boundary element (N/mm²)*
σ_{11}	−0.160	−0.200
σ_{12}	−2.300	−2.510

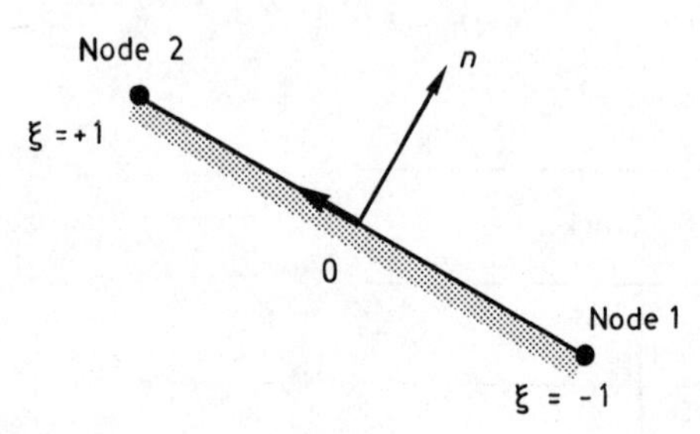

Fig. 5.9. Linear element

Let us now consider n linear boundary elements such that Equation (5.6) can be written,

$$c^i \mathbf{u}^i + \sum_{j=1}^{n} \left\{ \int_{\Gamma_j} \mathbf{p}^* \mathbf{\Phi}^T d\Gamma \right\} \mathbf{u}_j = \sum_{j=1}^{n} \left\{ \int_{\Gamma_j} \mathbf{u}^* \mathbf{\Phi}^T d\Gamma \right\} \mathbf{p}_j \qquad (5.24)$$

The integrals can be evaluated using numerical integration. Once this is done the element contribution can be assembled in much the same way as for the linear elements potential problem. The only difference is that now there are two unknowns per boundary node instead of one.

Example 5.4

In the following example we calculate the stress in a gear tooth[2]. The shaded part of the boundary is considered to be fixed and a P load normal to the surface and equal to 400 N/mm is applied, as shown in Figure 5.10. The tooth was analysed using plane strains and with different finite element and boundary element meshes.

For the finite element analysis the tooth is represented by 291 six-node isoparametric triangular elements with a total of 630 nodes (Figure 5.11(a)). For the boundary element analyses, linear, quadratic and cubic variation of displacement and traction over each element were tried. For the linear and quadratic case the boundary was subdivided into 33 elements (Figure

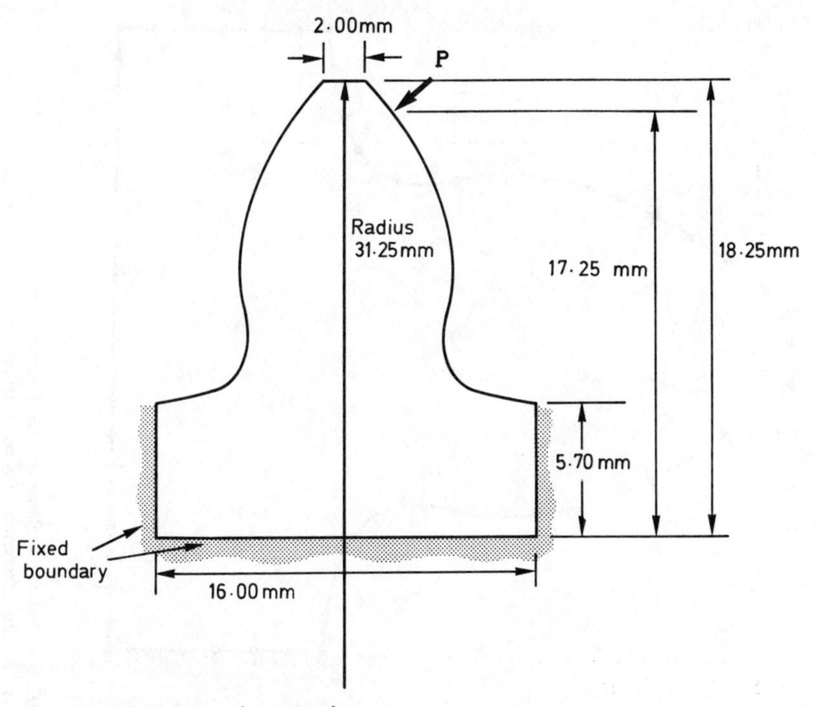

Fig. 5.10. Idealization of gear tooth

5.11(b)) and for the cubic case by 33 and also by 13 elements (Figure 5.11(c)). All runs were done on the same computer which means that the computer time presented in Table 5.4 are comparable. Notice that the quantity of data required by the finite element method is very much larger than that required by any of the boundary element programs. If a program for automatic generation of the finite element grid had been used the amount of data cards would be considerably reduced but some additional computer time would be required.

The variation of the calculated principal stress at the surface is shown in Figure 5.12. None of the solution gives sensible results near the line load so that part of the curve has not been plotted.

5.4 NON-HOMOGENEOUS CASE

Let us consider the case of a body divided in two different regions (Figure 5.13). The interface will be called Γ_I and we will assume two different fundamental solutions for regions 1 and 2. For Region 1 we have,

$\mathbf{p}^1$: Tractions on external surface of Region 1.

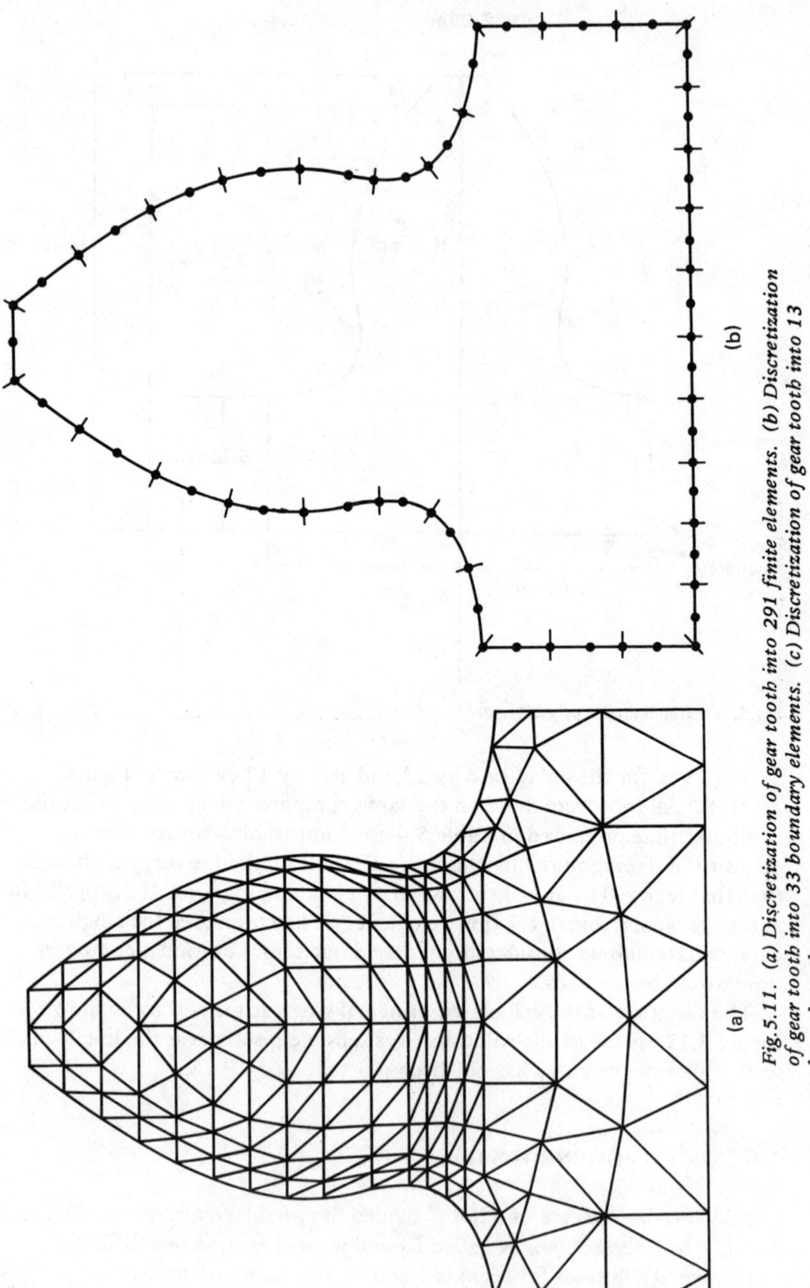

Fig.5.11. (a) Discretization of gear tooth into 291 finite elements. (b) Discretization of gear tooth into 33 boundary elements. (c) Discretization of gear tooth into 13 boundary elements

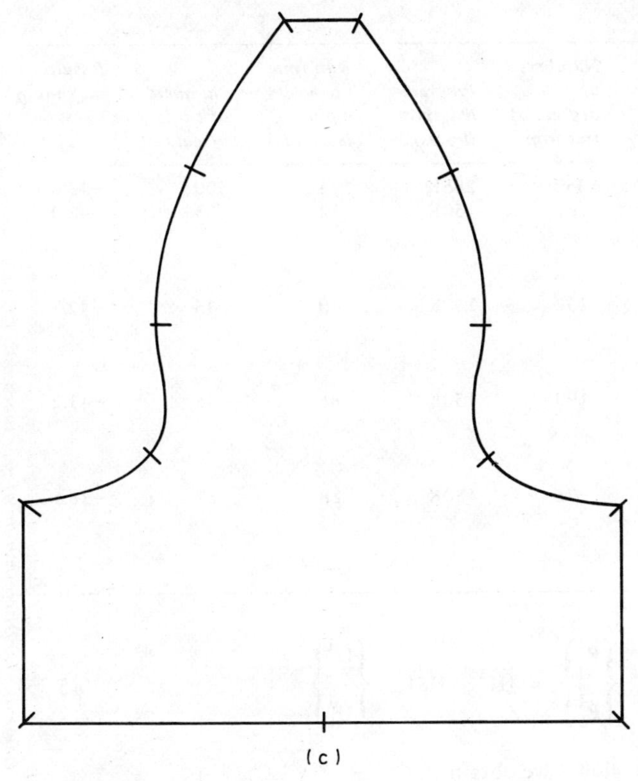

Fig. 5.11 (contd.)

$\mathbf{P}_I^1$: Tractions on interface Γ_I (considering that it belongs to Region 1)
$\mathbf{U}^1$: Displacements on the external surface of Region 1.
$\mathbf{U}_I^1$: Displacements on the interfacc Γ_I (considering that it belongs to Region 1).

For Region 2 of the boundary we have,

$\mathbf{P}^2$: Tractions on external surface of Region 2
$\mathbf{P}_I^2$: Tractions on interface Γ_I (considering that it belongs to Region 2)
$\mathbf{U}^2$: Displacements on the external surface of Region 2
$\mathbf{U}_I^2$: Displacements on the interface Γ_I (consider as belonging to Region 2).

The equations corresponding to Region 1 can be written as,

Table 5.4

Analysis	*Numbers of degrees of freedom*	*Program length (bytes)*	*Run time i.o. + c.p.u. (seconds)*	*Numbers of cards of data*	*Results maximum stress* (σ_{max})
Finite element	1260	286K	173	500	−43.0
Boundary element Linear 33 segments	66	150K	22	35	−42.0
Boundary element Quadratic 33 segments	132	150K	40	35	−42.7
Boundary element Cubic 33 segments	198	150K	86	35	−43.2
Boundary element Cubic 13 segments	78	150K	18	31	−38.1

$$[\mathbf{G}^1 \quad \mathbf{G}_I^1] \begin{Bmatrix} \mathbf{P}^1 \\ \mathbf{P}_I^1 \end{Bmatrix} = [\mathbf{H}^1 \quad \mathbf{H}_I^1] \begin{Bmatrix} \mathbf{U}^1 \\ \mathbf{U}_I^1 \end{Bmatrix} \tag{5.25}$$

Similarly for Region 2 we obtain

$$[\mathbf{G}^2 \quad \mathbf{G}_I^2] \begin{Bmatrix} \mathbf{P}^2 \\ \mathbf{P}_I^2 \end{Bmatrix} = [\mathbf{H}^2 \quad \mathbf{H}_I^2] \begin{Bmatrix} \mathbf{U}^2 \\ \mathbf{U}_I^2 \end{Bmatrix} \tag{5.26}$$

Let us call $\mathbf{P}_I$ and $\mathbf{U}_I$ the tractions and displacements on the interface Γ_I, defined using the following compatibility and equilibrium condition,

$$\mathbf{U}_I = \mathbf{U}_I^1 = \mathbf{U}_I^2 \tag{5.27}$$

$$\mathbf{P}_I = \mathbf{P}_I^1 = -\mathbf{P}_I^2 \tag{5.28}$$

Hence we can write Equations (5.25) and (5.26) as follows.

$$[\mathbf{G}^1 \quad \mathbf{G}_I^1 \quad -\mathbf{H}_I^1] \begin{Bmatrix} \mathbf{P}^1 \\ \mathbf{P}_I \\ \mathbf{U}_I \end{Bmatrix} = [\mathbf{H}^1] \{\mathbf{U}^1\} \tag{5.29}$$

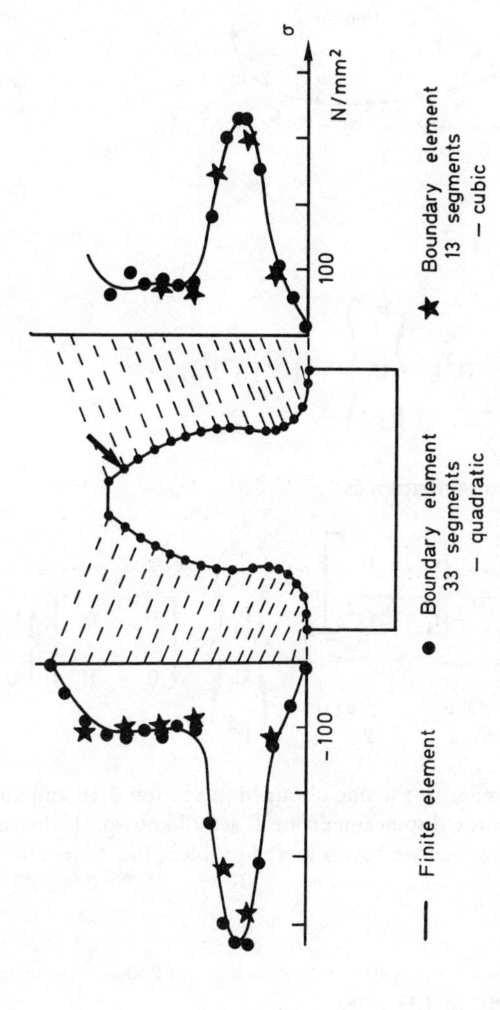

Fig.5.12. Variation of the principal stresses at the surface

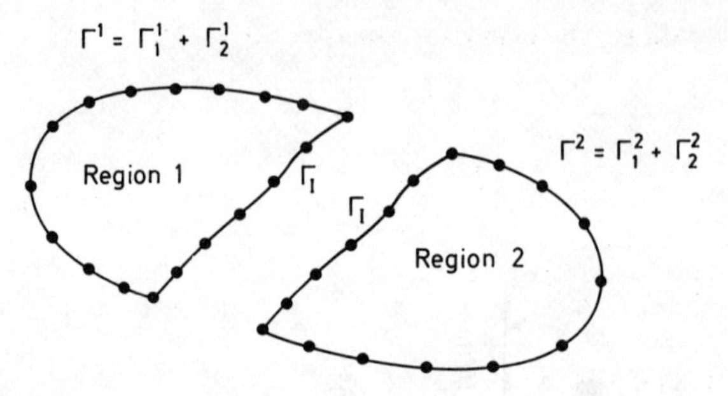

Fig. 5.13.

$$[-G_I^2 \quad -H_I^2 \quad G^2] \begin{Bmatrix} P_I \\ U_I \\ P^2 \end{Bmatrix} = [H^2]\ \{U^2\} \tag{5.30}$$

Both equations can be written as,

$$\underbrace{\begin{bmatrix} G^1 \\ 0 \end{bmatrix}}_{\text{Region 1}} \underbrace{\begin{matrix} G_I^1 & H^1 \\ -G_I^2 & -H_I^2 \end{matrix}}_{\text{Interface matrices}} \underbrace{\begin{matrix} 0 \\ G^2 \end{matrix}}_{\text{Region 2}} \begin{Bmatrix} P^1 \\ P_I \\ U_I \\ P^2 \end{Bmatrix} = \begin{bmatrix} H^1 & 0 \\ 0 & H^2 \end{bmatrix} \begin{Bmatrix} U^1 \\ U^2 \end{Bmatrix} \tag{5.31}$$

This result is similar to the one obtained in Section 3.10 and applies to a problem for which the displacements on Γ are all known. If instead of all displacements prescribed we have a mixed problem the system has to be reordered.

5.5 ANISOTROPIC SOLUTION

The solution for a two-dimensional anisotropic material has been presented in reference [3]. We will not repeat here the complete deduction but limit ourselves to showing the results relevant to somebody who wishes to program it.

First, let us start by defining the material properties such that,

$$\begin{Bmatrix} \epsilon_{11} \\ \epsilon_{22} \\ \epsilon_{12} \end{Bmatrix} = \begin{bmatrix} c_{11} & c_{12} & c_{13} \\ c_{12} & c_{22} & c_{23} \\ c_{13} & c_{23} & c_{33} \end{bmatrix} \begin{Bmatrix} \sigma_{11} \\ \sigma_{22} \\ \sigma_{12} \end{Bmatrix} \tag{5.32}$$

The c's are the material compliances. It can be shown[4] that for the material to be stable they have to satisfy the following fourth order equation,

$$c_{11}\mu^4 - 2c_{13}\mu^3 + (2c_{12} + c_{33})\mu^2 - 2c_{23}\mu + c_{22} = 0 \tag{5.33}$$

This is called the characteristic equation for the material and has four roots that are never real and are distinct if the material is not isotropic (for isotropic material $\mu = \pm i$).

$$\mu_j = a_j \pm ib_j, \quad j = 1, 2 \tag{5.34}$$

with $b_j > 0$ for thermodynamic considerations. μ_j define the characteristic directions for the material properties. For the $x_1 - x_2$ system they can be defined as a new system of coordinates $z_1 - z_2$, such that,

$$z_j = x_1 + \mu_j x_2, \quad j = 1, 2 \tag{5.35}$$

It is now more convenient to define two material constants r and q such that,

$$r_{11} = c_{11}\mu_1^2 + c_{12} - c_{13}\mu_1$$

$$r_{12} = c_{11}\mu_2^2 + c_{12} - c_{13}\mu_2$$

$$r_{21} = c_{12}\mu_1 + c_{22}/\mu_1 - c_{23}$$

$$r_{22} = c_{12}\mu_2 + c_{22}/\mu_2 - c_{23} \tag{5.36}$$

and,

$$q_{11} = \mu_1 \qquad q_{12} = \mu_2$$

$$q_{21} = -1 \qquad q_{22} = -1 \tag{5.37}$$

These constants are needed to define the displacements and tractions fundamental solution which are given by

$$u^*_{lk} = 2 \text{ Re } \{r_{k1}A_{l1} \ln(z_1) + r_{k2}A_{l2} \ln(z_2)\}$$

$$p^*_{lk} = 2 \text{ Re } \{q_{k1}(\mu_1 n_1 - n_2)A_{l1}/z_1 + q_{k2}(\mu_2 n_1 - n_2)A_{l2}/z_2\} \tag{5.38}$$

(Re{ } means real part of the complex number between brackets and Im { } will indicate its imaginary part). The direction cosines are

$$n_1 = \cos(n, z_1); \quad n_2 = \cos(n, z_2) \tag{5.39}$$

A_{lk} indicates a series of complex constants to be deduced from equilibrium and uniqueness of displacements around the singularity. We can first integrate the tractions on the boundary of a circular region around the singularity and equilibrate them with the internal load. This gives

$$\text{Im } \{A_{l1} + A_{l2}\} = -\frac{\Delta_{l2}}{4\pi}$$

$$\text{Im } \{\mu_1 A_{l1} + \mu_2 A_{l2}\} = +\frac{\Delta_{l1}}{4\pi}$$

$$l = 1, 2 \tag{5.40}$$

Integrating the strain along a circular path around the singularity we can obtain the displacements. The limits of the integrals are at the same point and so the integrals give a zero value, i.e.

$$\text{Im } \{r_{11}A_{l1} + r_{12}A_{l2}\} = 0$$

$$\text{Im } \{r_{21}A_{l1} + r_{22}A_{l2}\} = 0 \tag{5.41}$$

From Equations (5.40) and (5.41) we can deduce eight different constants (four for the real and four for the imaginary part of the A's).

References

1. Muskheliskvili, N. I., *Some Basic Problems on the Mathematical Theory of Elasticity*, Noordhoff Ltd., Holland (1953).
2. Lachat, J. C., "A Further Development of the Boundary Integral Technique for Elastostatics," Ph.D. Thesis, Southampton University (1975).
3. Cruse, T., U.S.A. Air Force Report, No.AFML-TR-71-268.
4. Lekhnitskii, S. G., *Theory of Elasticity of an Anisotropic Elastic Body*, Holden-Day Inc., San Francisco (1963).

6
Final remarks

6.1 RELATIONSHIP OF BOUNDARY ELEMENTS WITH FINITE ELEMENTS

The theoretical relationship between finite elements interpreted as a generalized Galerkin technique and boundary elements can be inferred from the basic concepts discussed in Chapters 1 and 2. This relationship is of practical importance if we intend to combine both methods for a particular problem. Although boundary elements tend to give much better answers than finite elements some problems with complex geometries or properties are simple to solve using finite elements. Furthermore it may be convenient to use finite elements in part of the domain and boundary elements in another part. Consider a problem such as the one of Figure 6.1 for which the properties of Region 1 are defined by a boundary element solution and the properties of Region 2 by finite elements. Let us assume that the problem is a potential problem for simplicity, although the same considerations will apply to an elasticity problem.

For the finite element solution one starts with

$$\int_{\Omega} (\nabla^2 u) w \, d\Omega = \int_{\Gamma_2} (q - \bar{q}) w \, d\Gamma \tag{6.1}$$

and assume that the function u satisfies the essential conditions $u = \bar{u}$ on Γ_1. In addition w is interpreted as being the arbitrary increment δu, such that w or δu and u are defined by the same set of functions.

Integrating (6.1) once by parts and taking into account that $w = \delta u$, we obtain,

$$\int_{\Omega} \frac{\partial u}{\partial x_k} \frac{\partial \delta u}{\partial x_k} \, d\Omega = \int_{\Gamma_2} \bar{q} \delta u \, d\Gamma \tag{6.2}$$

Functions are defined for u and δu over each element and a set of equations is finally obtained. These equations represent the properties of the system (stiffness in linear elastic problems or properties such as permeability,

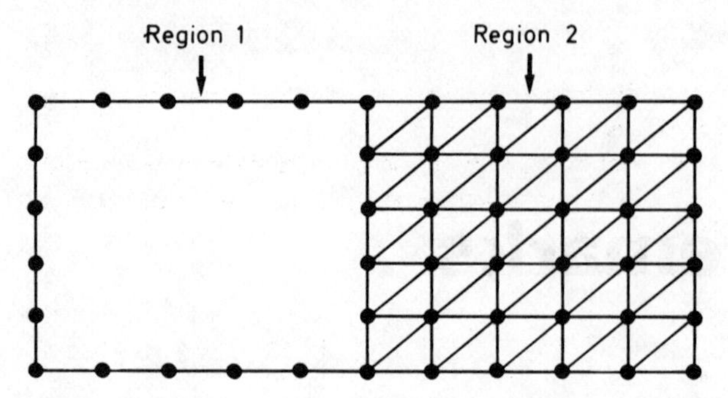

Fig.6.1. Combination of finite elements and boundary elements

heat conduction, etc. for potential problems). The final set of equations is usually written as,

$$\delta \mathbf{U}^T \mathbf{K} \mathbf{U} = \delta \mathbf{U}^T \mathbf{P} \tag{6.3}$$

or

$$\mathbf{K} \mathbf{U} = \mathbf{P}$$

Note that the unknowns are the potentials at the finite element nodes (or the displacements in elasticity). The actions **P** are said to be 'consistent' and they are obtained by calculating boundary integrals such as the ones on the right hand side of Equation (6.2). The elements of **P** are like actions concentrated at the nodes. They can be written as,

$$\delta \mathbf{U}^T \mathbf{P} = \sum_j \left(\int_{\Gamma_j} \bar{q} \delta u \, d\Gamma \right) \tag{6.4}$$

where j = number of elements on boundary Γ_2.

Note that the functions for u are defined on the boundary as,

$$u = \boldsymbol{\phi}^T \mathbf{u}_n \tag{6.5}$$

and the ones for q as

$$q = \boldsymbol{\psi}^T \mathbf{q}_n \tag{6.6}$$

where $\mathbf{u}_n$ and $\mathbf{q}_n$ are the node values of potentials and fluxes. The final expression for the consistent actions **P** can be deduced from

$$\delta \mathbf{U}^T \mathbf{P} = \Sigma \left[\delta \mathbf{u}_n \left\{ \int \phi \psi^T d\Gamma \right\} \mathbf{q}_n \right] = \delta \mathbf{U}^T \mathbf{M} \mathbf{Q} \tag{6.7}$$

From where, $\mathbf{P} = \mathbf{MQ}$ (6.8)

This equation represents the right hand side vector after assemblage. **Q** is a vector with the nodal values of q.

Let us next consider what happens with the boundary element method. Here we start with a weighted residual statement which takes into account the boundary conditions on Γ_1 as well as on Γ_2, i.e.

$$\int_{\Omega} (\nabla^2 u) w \, d\Omega = \int_{\Gamma_2} (q - \bar{q}) w \, d\Gamma + \int_{\Gamma_1} (u - \bar{u}) \frac{\partial w}{\partial n} \, d\Gamma$$

Integrating this equation by parts twice, we obtain

$$\int_{\Omega} u (\nabla^2 w) \, d\Omega = - \int_{\Gamma_2} \bar{q} w \, d\Gamma - \int_{\Gamma_1} q w \, d\Gamma + \int_{\Gamma_2} u \frac{\partial w}{\partial n} \, d\Gamma + \int_{\Gamma_1} \bar{u} \frac{\partial w}{\partial n} \, d\Gamma \tag{6.9}$$

Assuming that the w function is the fundamental solution u^* of the problem, one can write for a point 'i',

$$c^i u^i + \int_{\Gamma_2} u \frac{\partial u^*}{\partial n} \, d\Gamma + \int_{\Gamma_1} \bar{u} \frac{\partial u^*}{\partial n} \, d\Gamma = \int_{\Gamma_2} \bar{q} u^* \, d\Gamma + \int_{\Gamma_1} q u^* \, d\Gamma \tag{6.10}$$

This equation is integrated over all the boundary which produces a system of equations, i.e.

$$\mathbf{H}\,\mathbf{U} = \mathbf{G}\,\mathbf{Q} \tag{6.11}$$

We can interpret Equation (6.11) as the basic boundary element system before any boundary conditions have been applied. Note that Equation (6.11) includes function of the potential at the nodes (**U** vector) and the values of the fluxes q also at the nodes (**Q** vector). In order to work with Region 1 of Figure 6.1, as if it was a large finite element we have to build a finite element type matrix for Region 1 starting from Equation (6.11). First we invert **G**, i.e.

$$\{\mathbf{G}^{-1}\,\mathbf{H}\}\,\mathbf{U} = \mathbf{Q} \tag{6.12}$$

Then we can multiply both terms by the **M** distribution matrix defined in Equation (6.8). This gives,

$$(\mathbf{M}\,\mathbf{G}^{-1}\,\mathbf{H})\,\mathbf{U} = \mathbf{M}\,\mathbf{Q} = \mathbf{P} \tag{6.13}$$

This formula can be written as,

$$\mathbf{K}'\,\mathbf{U} = \mathbf{P} \tag{6.14}$$

The matrix $\mathbf{K}'$ is a finite element type matrix and should be symmetric from basic principles. In practice however $\mathbf{K}'$ is not symmetric due to the approximations involved. The simplest way of making $\mathbf{K}'$ a symmetric matrix is by using the method of least squares to minimize the error associated with $k'_{ij} \neq k'_{ji}$. If we call k_{ij} the coefficients of the new symmetric matrix $\mathbf{K}$ it is very easy to show that the least square error is a minimum when,

$$k_{ij} = \tfrac{1}{2}(k'_{ij} + k'_{ji}) \tag{6.15}$$

This means that the final and symmetric matrix $\mathbf{K}$ is

$$\mathbf{K} = \tfrac{1}{2}(\mathbf{K}' + \mathbf{K}'^{,T}) \tag{6.16}$$

Hence Equation (6.14) now becomes,

$$\mathbf{K}\,\mathbf{U} = \mathbf{P} \tag{6.17}$$

where $\mathbf{K}$ is the matrix for the Region 1 'element' shown in Figure 6.1 and has been found using boundary elements. The procedure is simple but expensive in computer time.

Effectively we have assumed that the Region 1 is an equivalent finite element and we can assemble the effective stiffness (Equation (6.16)) with those of the finite elements of Region 2. The final system is solved as a stiffness problem. Alternatively we can consider the two regions as if they were both boundary elements imposing compatibility and equilibrium conditions much in the same way as in Equations (3.10) and (5.2).

6.2 'INDIRECT' METHOD

The simplest boundary solution technique is the so-called 'indirect' method[1]. In its simplest form it starts with a solution that satisfies the governing equation in the domain but has some unknown coefficients which are then determined by enforcing the boundary conditions at a number of points i.e. for a potential problem

$$u = \overline{u}, \text{ on } \Gamma_1$$

$$q = \overline{q}, \text{ on } \Gamma_2 \tag{6.18}$$

The *indirect* method can be deduced from the *direct* one by 'considering the region Ω'', bounded by but 'exterior' to Ω, within which there are no sources and assume that u' is the solution to the reflex equation $\nabla^2 u' = 0$ within Ω'. Repetition of the above development, using u' in lieu of u leads to,

$$\int_\Gamma u^* q' d\Gamma + \int_\Gamma q^* u' d\Gamma = 0 \tag{6.19}$$

Note that because our point 'i' is now external to Ω' the corresponding first term in Equation (6.10) is now zero and the sign change in the right hand side terms is necessary so that q^* values in Equations (6.10) and (6.19) becomes identical despite the reversal of the unit normal direction, when considering Ω' rather than Ω. If we now specify u' to be that solution in Ω' which generates 'potentials around Γ identical to those in our initial problem, within Ω (i.e. on Γ, $u = u'$), one can add Equations (6.10) and (6.19) which produces for interior points,

$$u^i + \int_\Gamma u^*(q + q') d\Gamma = 0 \tag{6.20}$$

or

$$u^i = \int_\Gamma u^* \omega d\Gamma \tag{6.21}$$

Equation (6.21) is precisely the form of the *indirect* boundary element method statement for the problem. ω represents the initially unknown density of u^* or sources over Γ necessary to generate u^i via Equation (6.21).

We can also deduce the indirect approach in terms of dipoles if one assumes that $q + q' = 0$ instead of continuity of potentials. It gives,

$$u^i = \int_\Gamma \mu q^* d\Gamma \tag{6.22}$$

where the μ are called dipoles and give rise to the alternative indirect formulation. The formulation in terms of sources is usually preferred as it requires lower order derivatives of the u^* fundamental solution. The potentials and fluxes in the source approach are,

$$u^i = \int_\Gamma u^* \omega d\Gamma$$

$$q^i = \int_\Gamma q^* \omega d\Gamma \tag{6.23}$$

When the point 'i' is taken on the Γ boundary a singularity results such that for a smooth boundary,

$$u^i = \int_\Gamma u^* \omega \mathrm{d}\Gamma$$

$$q^i = -\tfrac{1}{2}\omega^i + \int_\Gamma q^* \omega \mathrm{d}\Gamma \tag{6.24}$$

The boundary can then be divided into a series of n elements for which n_1 are on Γ_1 and n_2 on Γ_2. By imposing boundary conditions (6.18) the values of the sources can be found. Once the values are known we can calculate the u's and q's at any point.

The method is very simple to apply and it gives reasonable results in many practical applications. It must be pointed out that the fundamental solution of several two-dimensional problems tend to infinity when the radius increases, instead of tending to zero due to its logarithmic character. In other words, the values of potential for instance, obtained for this case are relative and a constant C can be added to them. This constant ensures that the potential tends to zero when the radius increases. We can find the value of C by writing an extra equation as follows.

Note that when the radius increases the values of u_k^* tend to be the same for all the points on the boundary. This means that all the values of u_k^* for a large radius are approximately equal to

$$-\frac{1}{2\pi} \ln (r) \tag{6.25}$$

for the Laplace equation. Hence if we start with the discretized equation

$$u^i = \sum_{k=1}^{n} \int_{\Gamma_k} \omega_k u_k^* \mathrm{d}\Gamma + C \tag{6.26}$$

and divide all terms by $\ln (r)$, we obtain when $r \to \infty$.

$$\sum_{k=1}^{n_1} \omega_k = 0 \tag{6.27}$$

where the summation indicates all the unknown intensities ω_k on the boundary.

6.3 CONCLUSIONS

Boundary element solutions offer several important advantages over the 'domain' type solutions, such as finite element, finite differences, etc. These advantages are the possibility of defining only the surface of the body and the facility with which boundaries at infinity can be represented. The advantages are more marked in two- and three-dimensional continuum problems where concentration of stresses or fluxes occur. The method is also well suited to solve problems with boundaries at infinity such as those occurring in soil mechanics, hydraulics and other engineering disciplines, for which the classical 'domain' methods are obviously unsuitable.

The boundary method can be formulated in terms of influence functions and as such is frequently found in the literature under the general title of 'boundary integral method' but the weighted residual approach discussed here is more powerful and relates the method to more classical engineering techniques.

One of the more interesting features of the boundary element method is the simplicity of the input data required to run a problem. This contrasts with the large amount of data needed to run a finite element program. This is a very important point in practice as many man-hours are lost in preparing and checking finite element data.

The accuracy of boundary element solutions is generally greater than that of finite element techniques. Finite element results are usually accurate for the original variables under consideration (potential in certain field problems or displacements in stress analysis) but when these variables are differentiated (to obtain fluxes or stresses) the results are much less accurate and are usually discontinuous between elements. This problem can still be aggravated if regions of high flux or stress concentration exist in the continuum.

Boundary element matrices are usually fully populated which means that the method may be less efficient computationally than finite elements or others in certain application. To overcome this difficulty it has been suggested that the body ought to be divided into different regions, which gives a banded-type matrix. The subdivision of the body is also necessary for thin and long bodies or bodies which have dimensions of different order in different directions. In this respect it is generally inadvisable to exceed a ratio of 10.

Although it is beyond the scope of this book it is not difficult to extend the boundary method to study time dependent and non-linear problems. These effects can be incorporated in a similar way to the treatment of initial stress and strain fields.

In conclusion it can be asserted that the boundary element method presents definite advantages over 'domain-type' techniques and that these advantages are more evident for complex three-dimensional bodies and for problems with boundaries at infinity.

Reference

1. Brebbia, C. A. and Butterfield, R., 'The formal equivalence of the direct and indirect boundary element method', *Applied Mathematical Modelling,* 2, No.2 (1978)

Appendix

Numerical integration formulae

1. ONE-DIMENSIONAL GAUSSIAN QUADRATURE

$$I = \int_{-1}^{+1} f(\xi)\mathrm{d}\xi = \sum_{i=1}^{n} w_i f(\xi_i) + E_n \qquad \text{(A1)}$$

where w_i = weighting factor; ξ_i = coordinate of the ith integration point; n = total number of integration points; E_n = error = $0\ (\mathrm{d}^{2n}f/\mathrm{d}\xi^{2n})$

Table A1

n	i	ξ_i	w_i
1 (linear)	1	0	2
2 (cubic)	1	$+1/\sqrt{3}$	+1
	2	$-1/\sqrt{3}$	+1
3 (quintic)	1	0	8/9
	2	$+\sqrt{15}/5$	5/9
	3	$-\sqrt{15}/5$	5/9
4 (septimal)	1	+0.861 136 31	0.347 854 85
	2	−0.861 136 31	0.347 854 85
	3	+0.339 981 04	0.652 145 15
	4	−0.339 981 04	0.652 145 15

2. TWO- AND THREE-DIMENSIONAL GAUSSIAN QUADRATURE FOR RECTANGLES AND RECTANGULAR HEXAHEDRA

Two- and three-dimensional formulae are obtained by combining one-dimensional formulae according to

$$\int_{-1}^{+1}\int_{-1}^{+1} f(\xi, \eta)\mathrm{d}\xi \mathrm{d}\eta = \sum_{j=1}^{n}\sum_{i=1}^{n} w_i w_j f(\xi_i, \eta_j) \qquad \text{(A2)}$$

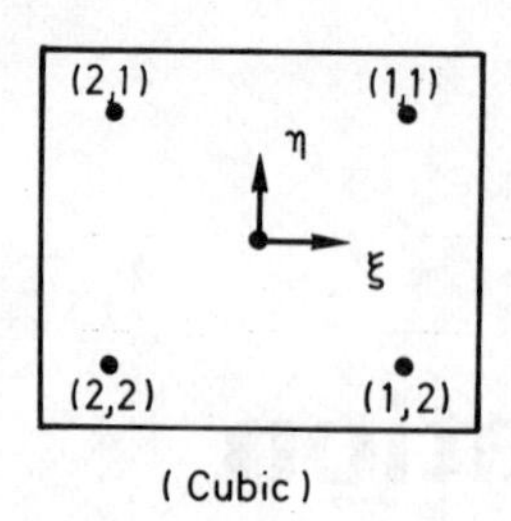

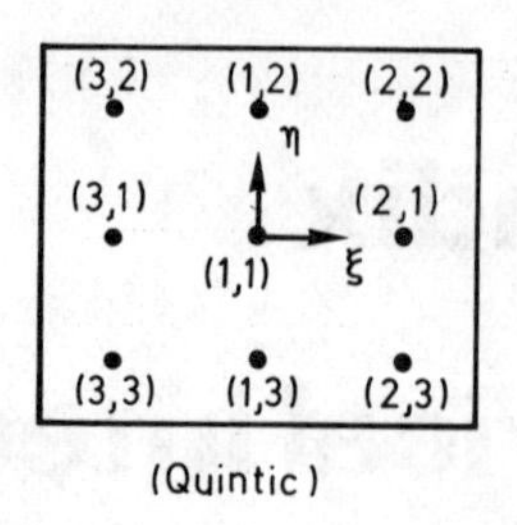

Fig.A1

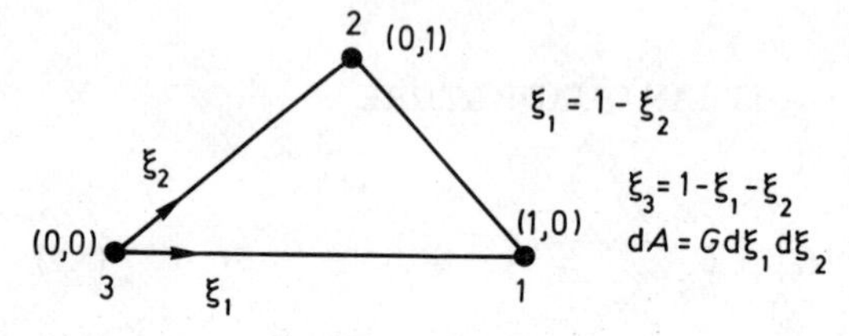

Fig.A2

$$\int_{-1}^{+1}\int_{-1}^{+1}\int_{-1}^{+1} f(\xi, \eta, \zeta)\mathrm{d}\xi\mathrm{d}\eta\mathrm{d}\zeta = \sum_{k=1}^{n}\sum_{j=1}^{n}\sum_{i=1}^{n} w_i w_j w_k f(\xi_i, \eta_j, \zeta_k) \qquad \text{(A3)}$$

where the weighting factors and integration point coordinates are listed in Table A1. The numbering scheme for the cubic and quintic two-dimensional rules are shown in Figure A1.

3. TRIANGULAR DOMAIN

$$I = \int_0^1 \left[\int_0^{1-\xi_2} f(\xi_1, \xi_2, \xi_3)\mathrm{d}\xi_1 \right] \mathrm{d}\xi_2 = \sum_{i=1}^{n} w_i f(\xi_1^i, \xi_2^i, \xi_3^i) \qquad \text{(A4)}$$

ξ_1, ξ_2, ξ_3 are triangular coordinates (see Figure A2). The exact integral for a combination of integral powers of the triangular coordinates is

$$I = \int_0^1 \left(\int_0^{1-\xi_2} \xi_1^i \xi_2^j \xi_3^k \mathrm{d}\xi_1 \right) \mathrm{d}\xi_2 = \frac{i!j!k!}{(i+j+k+2)!}$$

Quadrature formulae suggested by Hammer *et al.*[1] are listed in Table A2.

Table A2

n	i	ξ_1^i	ξ_2^i	ξ_3^i	w_i
1 (linear)	1	1/3	1/3	1/3	1
2	1	1/2	1/2	0	1/3
(quadratic)	2	0	1/2	1/2	1/3
	3	1/2	0	1/2	1/3
4 (cubic)	1	1/3	1/3	1/3	−9/16
	2	3/5	1/5	1/5	25/48
	3	1/5	3/5	1/5	25/48
	4	1/5	1/5	3/5	25/48
7 (quintic)	1	0.333 333 33	0.333 333 33	0.333 333 33	0.225 000 00
	2	0.797 426 99	0.101 286 51	0.101 286 51	0.125 929 18
	3	0.101 286 51	0.797 426 99	0.101 286 51	0.125 939 18
	4	0.101 286 51	0.101 286 51	0.797 426 99	0.125 939 18
	5	0.059 715 87	0.470 142 06	0.470 142 06	0.132 394 16
	6	0.470 142 06	0.059 715 87	0.470 142 06	0.132 394 16
	7	0.470 142 06	0.470 142 06	0.059 715 87	0.132 394 16

4. ONE-DIMENSIONAL LOGARITHMIC GAUSSIAN QUADRATURE FORMULAE

$$I = \int_0^1 \ln\left(\frac{1}{\xi}\right) f(\xi)\mathrm{d}\xi \simeq \sum_{i=1}^{n} w_i f(\xi_i) \tag{A5}$$

Quadrature formulae due to Stroud and Secrest[2] are listed in Table A3.

Table A3

n	i	ξ_i	w_i
2	1	0.112 008 20	0.718 539 31
	2	0.602 276 90	0.281 460 68
3	1	0.063 890 79	0.513 404 55
	2	0.368 997 06	0.391 280 04
	3	0.766 880 30	0.094 615 40
4	1	0.041 448 48	0.383 464 06
	2	0.245 274 91	0.386 875 31
	3	0.556 165 45	0.190 435 12
	4	0.848 982 39	0.039 225 48

n	j	ξ_i	w_i
5	1	0.029 134 47	0.297 893 47
	2	0.173 977 21	0.349 776 22
	3	0.411 702 52	0.234 488 29
	4	0.677 314 17	0.098 930 45
	5	0.894 771 36	0.018 911 55
6	1	0.021 634 00	0.238 763 66
	2	0.129 583 39	0.308 286 57
	3	0.314 020 44	0.245 317 42
	4	0.538 657 21	0.142 008 75
	5	0.756 915 33	0.055 454 62
	6	0.922 668 85	0.010 168 95
7	1	0.016 719 35	0.196 169 38
	2	0.100 185 67	0.270 302 64
	3	0.246 294 24	0.239 681 87
	4	0.433 463 49	0.165 775 77
	5	0.632 350 98	0.088 943 22
	6	0.811 118 62	0.033 194 30
	7	0.940 848 16	0.005 932 78

REFERENCES

1. Hammer, P. C., Marlove, O. J., and Stroud, A. H., "Numerical integration over simplex and cones", *Math. Tables and Other Aids to Computation,* Vol.10 (1956)
2. Stroud, A. H. and Secrest, D., *Gaussian Quadrature Formulas*, Prentice-Hall, New York (1966)

BIBLIOGRAPHY

Hammer, P. C., and Stroud, A. H., "Numerical evaluation of multiple integrals", *Math. Tables and Other Aids to Computation,* Vol.12 (1958)

Irons, B. M., "Engineering applications of numerical integration in stiffness method", *J. IAA,* 4, 2035–37 (1966)

Irons, B. M., "Economical computer techniques for numerically integrated finite elements", *Int. J. Numerical Meth. in Engng,* 1, 201–203 (1969)

Kopal, Z., *Numerical Analysis,* 2nd edn, Chapman & Hall (1961)

Miller, J. C. P., "Numerical quadrature over a rectangular domain in two or more dimensions", *Mathematics of Computation,* Vol.14, 13–20 (1960)

Index